G000057441

Estuaries

Estuaries: Monitoring and Modeling the Physical System

Jack Hardisty

Blackwell
Publishing

© 2007 by Jack Hardisty

BLACKWELL PUBLISHING
350 Main Street, Malden, MA 02148-5020, USA
9600 Garsington Road, Oxford OX4 2DQ, UK
550 Swanston Street, Carlton, Victoria 3053, Australia

The right of Jack Hardisty to be identified as the Author of this Work has been asserted in accordance with the UK Copyright, Designs, and Patents Act 1988.

All rights reserved. No part of this publication may be reproduced, stored in a retrieval system, or transmitted, in any form or by any means, electronic, mechanical, photocopying, recording or otherwise, except as permitted by the UK Copyright, Designs, and Patents Act 1988, without the prior permission of the publisher.

First published 2007 by Blackwell Publishing Ltd

1   2007

Library of Congress Cataloging-in-Publication Data

Hardisty, J. (Jack), 1955-
Estuaries : monitoring and modeling the physical system / Jack Hardisty
p. cm.
Includes bibliographical references.
ISBN-13: 978-1-4051-4642-5 (hardback : alk. paper)
ISBN-10: 1-4051-4642-7 (hardback : alk. paper)
1. Estuarine oceanography. 2. Estuarine oceanography–Mathematical models. I. Title.

GC97.H37 2007
551.46'18–dc22

2006029752

A catalogue record for this title is available from the British Library.

Set in 10.5/12.5 Meridien
by Newgen Imaging Systems (P) Ltd., Chennai, India
Printed and bound in Singapore
by Markono Print Media Pte Ltd

The publisher's policy is to use permanent paper from mills that operate a sustainable forestry policy, and which has been manufactured from pulp processed using acid-free and elementary chlorine-free practices. Furthermore, the publisher ensures that the text paper and cover board used have met acceptable environmental accreditation standards.

For further information on
Blackwell Publishing, visit our website:
www.blackwellpublishing.com

CONTENTS

PREFACE

This is the fourth book which the author has written, or almost written, in the field of environmental modeling. The first, which was a "how to" book entitled *Computerised Environmental Modelling: A Practical Introduction* with David Taylor and Sarah Metcalfe, appeared in 1993 and was reprinted within the year. Encouraged by this publishing success, a contract was offered for a similar "how to" book on beaches which I tried, without success, to write in 1995. The book was based upon an undergraduate course (and *vice versa*), but the course was deemed too difficult and the book drifted away. The course evolved into the more general area of coastal systems during 1996, and I tried to write the new book about the new course, but both book and course suffered from the weight of University administration and a number of difficult research projects. Finally, by 1996 the Internet had properly arrived and, with it, the possibility of using the World Wide Web as a research tool. My group engaged this exciting opportunity by developing operational estuary models and testing them with real time data on our web site. The result was exciting science and clearly of great practical interest to students seeking real world experience and to practitioners working in a wide range of coastal disciplines. A new, graduate module entitled "Estuaries: Monitoring, Modelling and Management" was introduced. In 2004 this was delivered online for the first time becoming simply "Estuaries" with streaming video and .wav files. In many ways this, the fourth draft but only the second to be published, is the book of that course.

This book rather than any of the others came to be written because of two complementary demands. First, it is a research monograph and arose out of work on the Natural Environment Research Council's (NERC's) Land Ocean Interaction Study (LOIS) program. The British Government, through its NERC, poured substantial investment into LOIS and many scientists from many disciplines attempted to coordinate their work toward a central research objective. This was the era of the "user community" and the concept of "information" or "decision support" systems gained considerable credibility within LOIS. Attempts were made to build such systems and, above all, Dr. Kevin Morris from the Plymouth Marine Laboratory shared the author's vision of combining data with operational forecasts (which, of course test our science) from within and without LOIS, and thus the concept of operational information systems was born. This book attempts to incorporate both the vision and the research results into the systems that are described here.

Second, it is a graduate and professional reference book and, in the United Kingdom at least, there is never enough time to cover all of the material which we would like in a single module. More importantly, students are always reluctant to pursue original source material and would rather work, quite reasonably, from a single unifying text. With this book the students of the University of Hull's module "16450 Monitoring and Modelling Estuarine Systems" will never again be able to claim that they were unable to locate the reading material. It is all here.

This is also a dynamic book with many of the models, calculations, diagrams, and internet references available and updated through the book's web site at www.blackwellpublishing.com/hardisty

Finally, grateful thanks must be extended to the very many people who have helped in the development of the ideas described here but, of course, the errors and omissions are mine alone.

Jack Hardisty

Professor of Environmental Physics, The University of Hull

j.hardisty@hull.ac.uk

Microsoft product screen shots reprinted with permission from Microsoft Corporation.

ESTUARINE TOOLBOXES

The Estuarine Toolboxes are available from www.blackwellpublishing.com/hardisty

DYNAMIC INTERNET REFERENCES

The following internet references were used for the images described. Updated internet references, where appropriate, are available from www.blackwellpublishing.com/hardisty

DIR 1.1   The Humber Estuary: glcf.umiacs.umd.edu/index.shtml

DIR 1.2   Eccentricity, tilt, and precession of the earth's orbit: after www.homepage.montana.edu/~geol445/hyperglac

DIR 1.3   Precession, tilt, and eccentricity combine to force solar heating and result in glacial advances and retreats: after en.wikipedia.org/wiki/Milankovitch_cycles

DIR 1.4   Sea-level curves for various sites around the United Kingdom: after home. hiroshima-u.ac.jp/er/Resources/Image247.gif

DIR 1.5   Tides in the Humber Estuary for October 19, 2004 for (from top) Spurn Head, Immingham, Humber Bridge, and Goole: from easytide.ukho.gov.uk/EasyTide

DIR 2.1   Tide pole at Anchorage, Alaska: from www.co-ops.nos.noaa.gov/about2.html

DIR2.2   Deep water pressure transducer gauge: www.valeport.co.uk

DIR 2.3   Acoustic tide gauge installed at Settlement Point, Bahamas: sealevel.jpl.nasa.gov/science/invest-merrifield-fig2.htm

DIR 2.4   British Oceanographic Data Centre: www.pol.ac.uk/ntslf/

DIR 2.5   NOAA's Tides Online: tidesonline.nos.noaa.gov/geographic.html

DIR 2.6   ACCLAIM: www.pol.ac.uk/psmsl/programmes/acclaim.info.html

DIR 2.7   Valeport Braystoke impellor flow meter: www.valport.co.uk

DIR 2.8   (a) Travel Time acoustic current meter and (b) Döppler shift acoustic current meter: www.sontek.co

DIR 2.9   A range of electromagnetic current meters: www.valeport.co.uk

DIR 2.10   The North American PORTS system: co-ops.nos.noaa.gov/d_ports.html

DIR 2.11   Traditional oceanographic reversing thermometer: www.photolib.noaa.gov/ships/ship3154.htm

DIR 2.12   Seabird Electronics oceanographic temperature sensor for use in estuarine waters: www.seabird.com

Part I

EVOLUTION AND MONITORING

1

INTRODUCTION TO ESTUARINE SYSTEMS

Contents

1.1 INTRODUCTION

The term estuary is derived from the Latin word "*aestus*" meaning tide and refers to a tongue of the sea reaching inland. Estuaries are formed by sea-level rise following a glaciation or ice age (Woodroffe, 2003), and represent the complex nonlinear interaction of tides, currents, salt, water, and sediment. In this book, we attempt to combine the latest theoretical and empirical results to build and to test a new integrated and transparent model which simulates these estuarine processes. In this chapter we examine the longer-term climate changes which lead to the sea-level rise and estuarine formation and describe the resulting, modern estuarine processes and some estuarine classifications.

There are a number of excellent text books and journals dedicated to different aspects of estuarine science and management. As a general introduction, Brown (1999) provides a readable, largely qualitative description of estuarine processes. Quantitative rigor is provided by Dyer's books (1986 and 1997) and specialist publications in the various disciplines covered here: oceanography, sedimentary geology, meteorology, engineering, physical geography, and the environmental sciences.

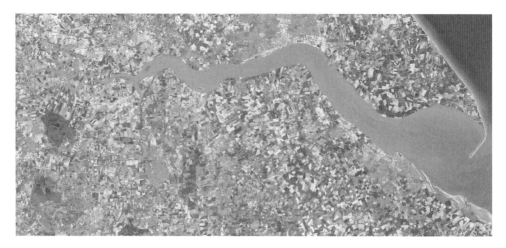

FIGURE 1.1 Landsat image of the Humber Estuary from Spurn Head (in the east on the right of the image) to the headwaters (in the west). Image cropped from a LANDSAT 7 satellite data file code p202r023_7t20010512 supplied by the Global Land Cover Facility, University of Maryland, College Park, USA. URL (DIR 1.1).

Although this book deals with a generalized, virtual estuary, it is based on research work in the British Natural Environment Research Council's Land Ocean Interaction Study (LOIS) on the author's local waterway called the Humber Estuary (Figure 1.1). The Humber is ideal for these purposes. It has a catchment of some 25,000 km^2, with a large tidal range and strong currents, and is responsible for the interchange of millions of tonnes of sediment with the North Sea each year. The Humber also has great socio-economic importance with a long history of marine and maritime trade and traffic and major port and petrochemical complexes. It is also typical of the world's estuaries in that it is, in geological terms, young and dynamic being a product of sea-level rise associated with the retreat of ice following the last global glaciation.

We begin with a qualitative model that describes estuarine processes in five stages:

1 **Bathymetry**: A river mouth is flooded with ocean water during sea-level rise after a glacial period generating a basic three-dimensional shape.

2 **Tides**: The ocean is tidal. At high tide, seawater flows into the river mouth making the estuary more saline whilst at low tide the water returns to the sea making the estuary less saline.

3 **Currents**: The inflow, outflow, and mixing of the ocean water with the land drainage generates freshwater and tidal currents within the estuary.

4 **Temperature and salinity**: The tidal currents transport heat and salt around the estuary through the processes of advection and diffusion.

5 **Particulates**: Solid particles are also eroded, transported, and deposited, so that the bathymetry changes and in turn influences the tides, currents, and transport processes.

The processes are sequentially detailed in later sections of this book. For example, the theory and modeling of estuarine bathymetry is covered in Part II. Tides, currents, temperature and salinity, and particulate matter are covered in Parts III, IV, V, and VI respectively.

1.2 ORIGINS, CLIMATE, AND ICE AGES

The geological history of coastal and estuarine environments is best understood in terms of global evolution due to the processes of plate tectonics and climate change that have been continuing since the planet first solidified some 4,600 myBP (million years before present). Geologists divide these vast intervals of time into four distinct eras and into periods and epochs as shown in Table 1.1. Archaeologists have divided the Pleistocene and Holocene Epochs into a series of ages, related to the utilization of tools by mankind.

It is clear from the geological record that there are more or less regular cycles in the Earth's climate within which colder conditions lead to the onset of ice ages during which sea-levels fall by tens to hundreds of metres. Following amelioration and the retreat of the ice, sea-level rises and the associated marine transgressions flood river valleys and form estuaries on a global scale. It is also now clear that it is the relatively increased solar radiation and summer warming and melting of the ice associated with these cycles which is one of the most important driving mechanisms. The solar radiation levels are, in turn, controlled by the precession and tilt of the Earth's axis and the eccentricity of Earth's orbit around the Sun as shown in Figure 1.2. It is now known that changes in these orbital parameters are responsible for global climate change and, in particular, for the advance and retreat of a series of ice ages throughout geological time and, ultimately, for the formation of estuaries.

1 **Precession** is the Earth's axis' slow rotation as it spins. This top-like wobble, or precession, has a periodicity of about 23,000 years.
2 **Tilt** varies between about 21° and 24° over a period of about 41,000 years and the direction changes over two cycles of 19,000 and 23,000 years. Today the Earth's axial tilt is about 23.5°, which largely accounts for our seasons. Because of the periodic variations of this angle the severity of the Earth's seasons changes. With less axial tilt, the Sun's solar radiation is more evenly distributed between winter and summer. Less tilt also increases the difference in radiation receipts between the equatorial and polar regions.
3 **Eccentricity** of the orbit varies over a period of about 100,000 years. At present the orbital eccentricity is nearly at the minimum of its cycle with a difference of only about 3% between aphelion (farthest point) and perihelion (closest point) so that Earth receives about 6% more solar energy in January than in July. When the Earth's orbit is most elliptical the amount of solar energy received at the perihelion would be in the range of 20–30% more than at aphelion.

There have been eight large glacial buildups over the past 800,000 years, each coinciding with a minimum eccentricity (Figure 1.3) and associated sea-level changes.

1.3 WEB SITE SYSTEMS

This book is supported by a series of web pages accessed through the publisher's site at **www.blackwellpublishing.com/hardisty** where three groups of tools may be accessed:

1 **Dynamic Internet References**. There are a number of online references throughout this book to Internet sites, which are kept up to date with Dynamic Internet References

TABLE 1.1 The divisions of geological time.

Eras	Periods	Epochs	Ages
Cenozoic 63 myBP	Quaternary 2 myBP	Holocene 10,000 BP	Roman 43 AD
			Iron 600 BC Late, middle and early
			Bronze 2000 BC Late, middle and early
			Neolithic (New Stone) 3,500 BC late and early
			Mesolithic 10,000 BC Late middle and early
		Pleistocene	Paleolithic 1,000,000 (1 my) Upper, middle and lower
	Neogene 26 myBP	Pliocene	
		Miocene	
	Palaeogene 63 myBP	Oligocene	
Mesozoic 225 myBP		Eocene	
		Palaeocene	
	Cretaceous 136 myBP	Land epoch in Britain and transgression *South America separates* *from South Africa*	
	Jurassic 195 myBP	Downwarping in Britain *initiation of* *North Atlantic*	
	Triassic 225 myBP	Erosion, infilling, and marine transgression	
Paleozoic 570 myBP	Permian 280 myBP	Pennine uplift and lowering of sea-level	
	Carboniferous 345 myBP	Marine transgression and complex elevation	
	Devonian 410 myBP	Folding, warping, and local sedimentations	
	Silurian 440 myBP	Final infilling of the Caledonian basins	
	Ordovician 530 myBP	Folding, downwarping, and metamorphism	
	Cambrian 570 myBP	Seafloor spreading and sedimentation	
Pre-Cambrian **From 4,600 myBP**		Lewisian and Torridonian rocks of northwest Scotland date from Pre-Cambrian times	

All dates mark the lower boundary of the interval.

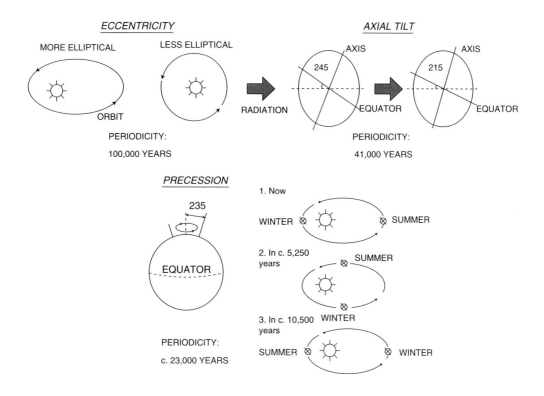

FIGURE 1.2 Eccentricity, tilt, and precession of the Earth's orbit (after DIR 1.2).

(DIR) on the book's web site. Thus, selecting DIR 1.3 from the caption in Figure 1.3 automatically connects the browser to the latest version of the diagram.

2 **Estuarine Toolboxes**. Second, there is an Excel workbook on the site which contains a number of useful implementations of the equations and formulae which are introduced in the text to represent the estuarine processes. These are called the Estuarine Toolboxes and are utilized in later analysis. For example, we frequently use the cyclic trigonometric function represented by the circular sinusoid to describe estuarine processes in this book:

$$\sin \alpha = \frac{e^{i\alpha} - e^{-i\alpha}}{2} \qquad \text{Eq. 1.1}$$

where α is the angle. For example, the precession, $P(t)$, in the Milankovitch cycle might be represented by

$$P(t) = P_1 \sin \left[\frac{2\pi\,(t + \rho_1)}{T_1} \right] + P_2 \sin \left[\frac{2\pi(t + \rho_2)}{T_2} \right]$$

$$+ P_3 \sin \left[\frac{2\pi(t + \rho_3)}{T_3} \right] \qquad \text{Eq. 1.2}$$

where $P_1, P_2,$ and P_3 are the amplitudes of components with periods $T_1, T_2,$ and T_3, and phase differences of $\rho_1, \rho_2,$ and ρ_3, respectively. The results are shown in Figure 1.4 and implemented in Estuarine Toolbox 1 for three components of 19, 22, and 24 ka.

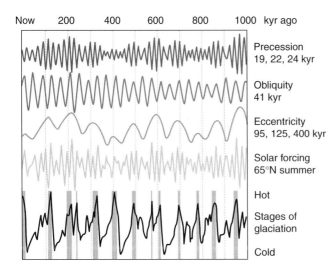

FIGURE 1.3   Precession, tilt, and eccentricity combine to force solar heating and result in glacial advances and retreats (after DIR 1.3).

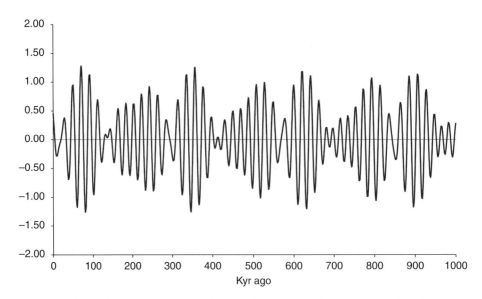

FIGURE 1.4   Simulating the precession cycles shown in Figure 1.3 with Estuarine Toolbox 1.

3 Estuarine Models. Each of the modeling chapters in this book (Chapters 4, 6, 8, 10, and 12) explains how the processes are incorporated into the overall spreadsheet model. The Estuarine Models section of the web site includes examples of this coding as it develops through the book.

1.4   SEA-LEVEL RISE AND ESTUARIES

Field evidence (e.g. Summerfield, 1991) demonstrates that sea-level responds to the advance and the retreat of the ice sheets through a combination of two processes:

1 **Eustatic sea-level change**: global changes in sea-level connected with the abstraction and return of water from the oceans as the ice freezes and melts which results in a sea-level fall during the ice age.
2 **Isostatic sea-level change**: localized crustal deformation associated with the loading and unloading of glacial ice which results in sea-level rise during the ice age and a sea-level fall following the retreat of the ice.

As a consequence we often speak of relative sea-level change – the height of sea relative to land. Sea-level change is often difficult to ascertain due to complex changes in both oceanic water volume and isostasy. At 18,000 BP sufficient water had been removed and locked up in glaciers such that sea-level had fallen by 130 m. Yet eustatic lowering was accompanied by isostatic depression in Scandinavia, Britain, and Canada. Following wastage of the northern hemisphere ice sheets, global sea-levels rose steadily while thinning resulted in rapid isostatic recovery. Isostatic changes have been documented in Scandinavia (700 m of uplift in Holocene), eastern Scotland (250 m of uplift since deglaciation), and Laurentia (900 m of uplift since deglaciation). In the early Holocene, eustatic sea-level rise ($0.01 \, \mathrm{m \, yr^{-1}}$) began to exceed isostatic rebound resulting in a marine transgression around the coastline of Scotland between about 8,500 and 6,500 thousand years ago (Figure 1.5).

The paleography of the Humber region during and following the last glaciation is shown in Figure 1.6. Before the ice age, the North Sea coastline lay along the eastern flanks of the Yorkshire and Lincolnshire Wolds. During the period after the retreat of the ice, but before the sea-level rose finally, East Yorkshire was part of the Dogger Bank land complex. The two River Hulls drained into the Humber and flowed east and north toward the coastline.

A detailed coring analysis of the sediments within the Humber basin was undertaken as one element of the Natural Environment Research Council's (NERC) LOIS (Shennan and Andrews, 2000). The results from Metcalfe et al. (2000) are shown in Figure 1.7. For these reconstructions, five major environments were distinguished:

1 Upland;
2 Raised bog;
3 Eutrophoic wetland and freshwater aquatic;
4 Intertidal; and
5 Estuarine subtidal.

8,000 calibrated years BP reconstruction

Estuarine subtidal and intertidal environments were restricted to the outer estuary (Figure 1.7a) when sea-level was about 17 m below present. The mainly marine, Garthorpe suite described by Rees et al. (2000) represents these environments with a widespread occurrence of eutrophic wetland with mosaic of oak-hazel fenwood, open standing water, and sedge fen. The main channel followed a meandering course around the line of the present estuary. Other boreholes (e.g., Van de Noort and Ellis, 1997) have located older, deeply incised channels of the main rivers. There is also some evidence of at least temporary incursions of salt water beyond the tidal limits which could be related to the effects of the tsunami recorded widely in northeast Scotland.

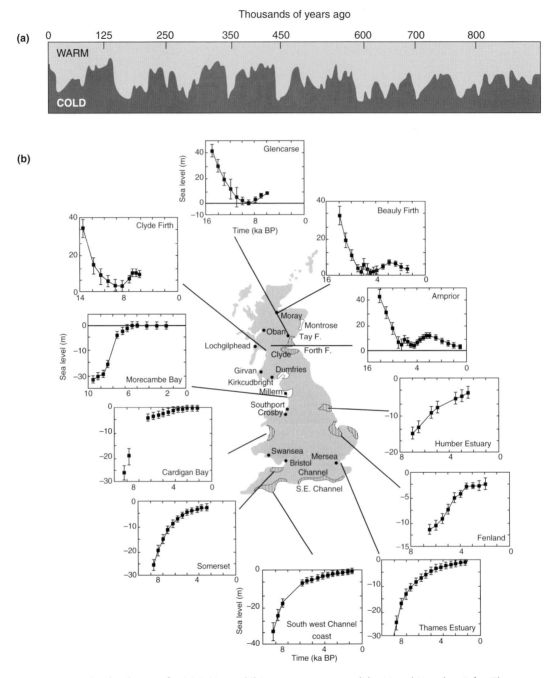

FIGURE 1.5 Sea-level curves for (a) 1 Ma and (b) various sites around the United Kingdom (after Shennan et al., 2002) (after DIR 1.4).

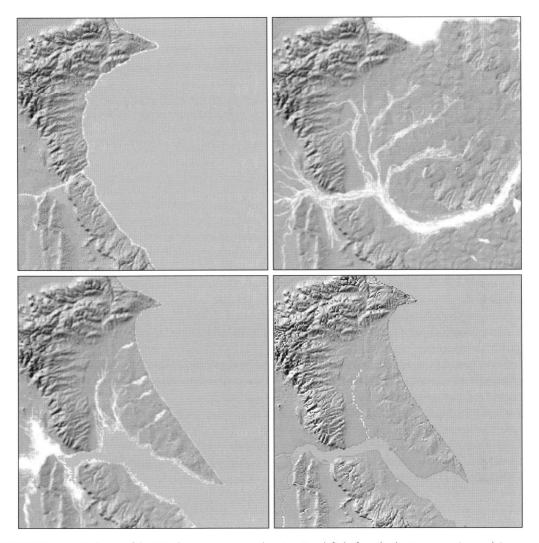

FIGURE 1.6 Evolution of the Humber environment showing (top left) before the last ice age; (top right) immediately following the retreat of the ice; (bottom left) approximately 2,000 BP; and (bottom right) the present day coastal alignment.

7,000 calibrated years BP reconstruction

Sea-level is about 10 m below the present level and the facies show the transgression of intertidal sediments across the eutrophic wetland (Figure 1.7b). The intertidal limit occurs at the mouth of the Ancholme and the inner estuary remains dominated by eutrophic wetland of the Newland and Butterwick suites (Rees et al., 2000).

6,000 calibrated years BP reconstruction

The transgression continues up-estuary (Figure 1.7c) with intertidal environments encroaching into the lower reaches of

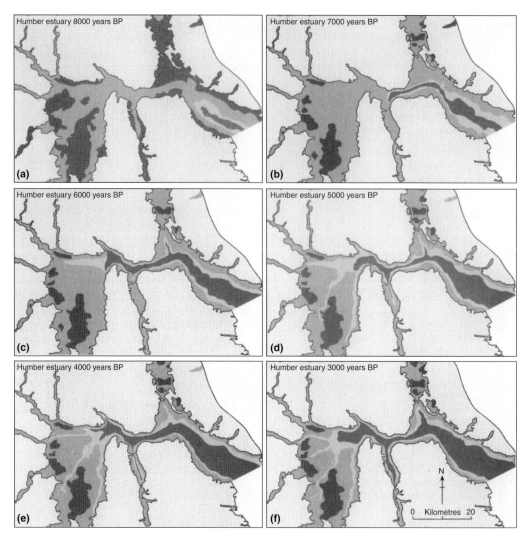

FIGURE 1.7 Holocene evolution of the Humber Estuary. (a) 8,000 calibrated years BP reconstruction, (b) 7 ka BP, (c) 6 ka BP, (d) 5 ka BP, (e) 4 ka BP, and (f) 3 ka BP (after Metcalfe et al., 2000).

the Trent, Hull, Ouse, and Foulness Valleys (Long et al., 1998).

5,000 calibrated years BP reconstruction

The reconstruction (Figure 1.7d) shows that intertidal sedimentation occurs in all of the valleys draining into the estuary but the

data are insufficient to accurately define the narrow tidal reaches of the individual rivers (Metcalfe et al., 2000).

4,000 calibrated years BP reconstruction

Intertidal environments continue to expand up the valley systems as shown in Figure 1.7e, but there are again insufficient data to enable a more detailed reconstruction (Metcalfe et al., 2000).

3,000 calibrated years BP reconstruction

This reconstruction (Figure 1.7f) represents the probable maximum extent of the intertidal intrusion. There is some evidence of a regression after this stage (Long et al., 1998).

1.5 BATHYMETRY

Bathymetry is the three-dimensional shape of the estuary. Estuaries that are used for navigation and shipping are routinely surveyed and bathymetric charts are drawn showing water depths, bathymetric contours, and, usually, a wide range of additional information about tides and currents. Modern techniques for bathymetric surveying are described in Chapter 2. Theoretical considerations are examined in Chapter 3, and basic models of estuarine bathymetry are developed and tested in Chapter 4. The bathymetry of the Humber Estuary is shown in Figure 1.8 and Gameson (1982) provides a general description which accords with the three zones of Dalrymple et al. (1992):

1 **Upper Humber**. The majority of the upper Humber dries out at low water where the exposed shoals are composed mostly of fine sand with some silt, whereas the littoral zone is covered in fine mud. Examination of historical charts shows that before the construction of the training walls at Trent Falls in the mid-1930s, the channel oscillated from side to side depending upon the relative strength of the flows from the Ouse and the Trent. However, more recently, the reach has stabilized with the main channel on the south side immediately downstream of Trent Falls. Second, there is a marked variation in channel and shoal locations in the vicinity of Read's Island. The low-water channel migrates from the northern shore to south of Read's Island until prolonged and high river flows force a new channel through the Winteringham Middle Sand to develop the Redcliffe Channel. The Ancholme Channel then appears to decay as the Redcliffe migrates southward to run, eventually, off the northern side of Read's Island until the Ancholme line is once more occupied. The process appears to have occurred three times during the twentieth century.

2 **Middle Humber**. Between Hessle and Immingham there is also an apparent preferred configuration. The main channel crosses from the southern shore at the Humber Bridge, to run close to the Hull waterfront. Seaward, the reach is dominated by Skitter and Foul Holme Sands, with the channel crossing again to the south bank to provide the deep water moorings for the Immingham port complex. A well-developed Barton Ness Sand appears to supply material to Skitter Sand which then extends further seaward.

3 **Lower Humber**. The lower Humber is dominated by a three-channel system at the mouth: the northern Hawke Channel leads upstream through the dredged Sunk Channel, whereas the central Bull Channel and southern Haile Channel converge onto Middle Sand. The littoral zones are relatively stable, with small seasonal variations in the elevation of Spurn Bight. The southern shore is subject to wave attack which is likely to redistribute the surface sediments consisting mainly of fine sand.

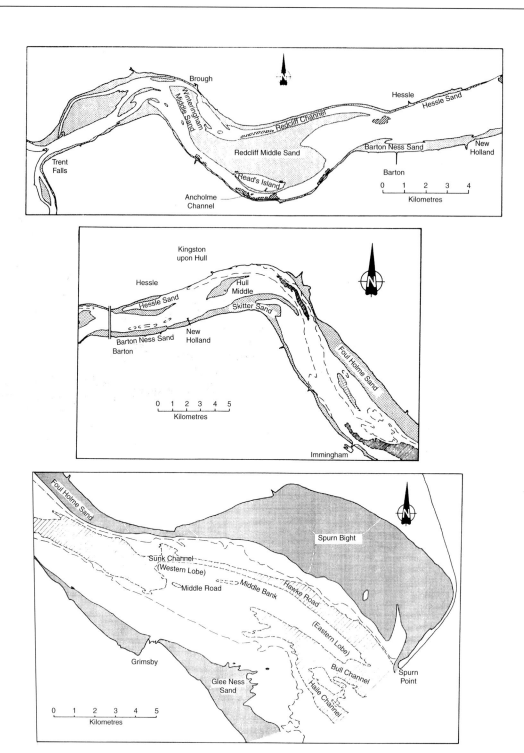

FIGURE 1.8 The bathymetry of the Humber Estuary (after Gameson, 1982) with (top) Upper Humber, (middle) Middle Humber, and (bottom) Lower Humber. Stippled areas show banks exposed at Chart Datum (approximately, low-water Spring Tide).

1.6 TIDES

Tides are the regular rising and falling of the waters of the world's oceans due to the gravitational attraction of the Moon and, to a lesser extent, of the Sun (Pugh, 2004). We cover the mechanics of tide generation and propagation in detail in Chapter 5 but require here a basic introduction. Tidal measurements (Chapter 2) are plotted as time series of water depths. For example, Figure 1.9 shows tides at four locations along the Humber Estuary for October 18, 2004. The North Sea tide begins to rise at Spurn Head at about 0130 and the progress up the estuary can be followed with high water occurring at Spurn at about 0745 and reaching Goole at 0930. Water levels then fall and the afternoon tide repeats the sequence. Four terms can be defined from this sequence:

1 **High water** is the maximum water depth (or crest of the tidal wave) with respect to the datum.
2 **Low water** is the minimum water depth (or trough of the tidal wave) with respect to the datum.
3 **Tidal range** is the vertical distance between high water and low water.
4 **Tidal period** is the time interval between the occurrence of sequential high waters (or low waters or any other corresponding point on the tidal wave).

We can immediately begin to quantify these terms so that, from Figure 1.9,

1 High water at Spurn Head is 6.90 m Chart Datum.
2 Low water at Spurn Head is 1.50 m Chart Datum.
3 The tidal range is 5.4 m.
4 Morning high water occurred at approximately 0743 and the evening high water

occurred at approximately 2005 at Spurn Head and thus the tidal period was approximately 12 hr 22 min (12.3 hr).

1.7 CURRENTS

Currents are generated by both the rising and falling of the tides and by the input of freshwater from the catchment. We cover the mechanics of current generation in detail in Chapter 7 but require here a basic introduction. Traditionally, tidal measurements are plotted as tidal diamonds on navigational charts. For example, Figure 1.10 shows currents along the Humber for October 18, 2004. The tide begins to flow landward at Spurn Head at about 0130 with maximum flows about 3 hours later, before slackening off to a still stand at high water. The water then flows seawards until the afternoon tide repeats the sequence. Three terms can be defined from this sequence:

1 **Flood tide** is the current in the direction of tidal propagation (into the estuary).
2 **Ebb tide** is the current in the opposite direction to tidal propagation (out of the estuary).
3 **Maximum flood and ebb** usually occur at mid tide in estuary environments.

We can immediately begin to quantify these terms:

1 Maximum flood at Spurn Head was about 2.7 knots (about $1.35\,\mathrm{m\,s^{-1}}$).
2 Maximum ebb at Spurn Head was about 3.2 knots (about $1.60\,\mathrm{m\,s^{-1}}$).

1.8 TEMPERATURE AND SALINITY

Although water properties are usually taken to include the very wide range of chemical

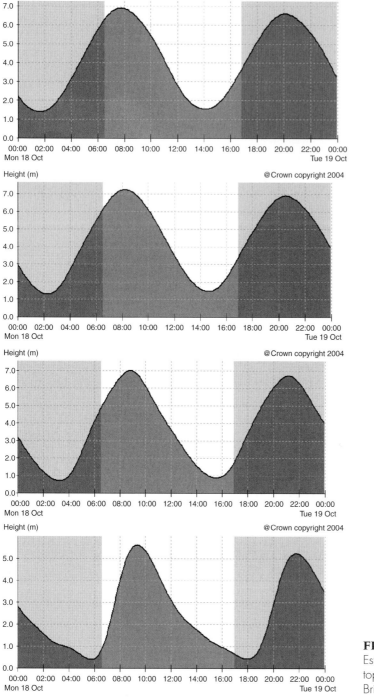

FIGURE 1.9   Tides in the Humber Estuary for October 19, 2004 for (from top) Spurn Head, Immingham, Humber Bridge, and Goole (from DIR 1.5).

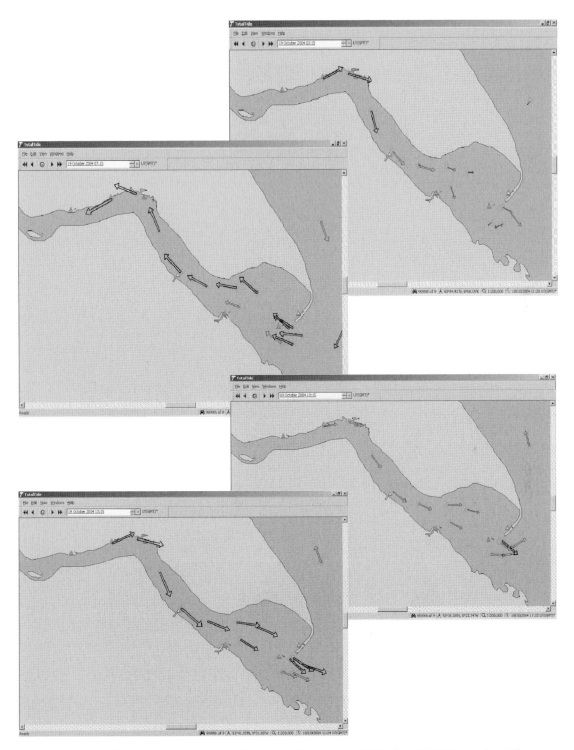

FIGURE 1.10   Tidal currents in the Humber Estuary for 18 October, 2004 from UK Hydrographic Office (2004).

and biological substances found in estuarine waters, we shall confine our attention here to three parameters: (i) the temperature, (ii) the salinity, and (iii) the suspended particulate matter (SPM). Details of water properties and the theoretical aspects of advection and diffusion are given in later chapters of this book. For the present purposes, we require an introduction that will again be illustrated with reference to the Humber.

Temperature

The temperature of estuarine waters varies on daily and seasonal time scales and also spatially, depending upon the relative temperatures of the tidal and freshwater inputs. In general, in temperate latitudes, the freshwater is colder than the seawater in winter so that there is a positive temperature gradient in a seaward direction and toward high water. Conversely, the freshwater is warmer than the seawater in summer, so that the gradients operate in the opposite direction. An example of the temperature close to the surface along the Humber Estuary is given in Figure 1.11a, taken from Gameson (1982).

Salinity

The sea is salty because water is a very good solvent. Solids are dissolved in seawater and the most commonly occurring is common salt, sodium chloride. Open ocean waters have a salinity of 34–36 g kg^{-1}, and there is then a gradient in salinity increasing from low and brackish values in the head waters toward the open sea. The transfer of salt is a diffusive process (Chapter 9) and thus salinity is usually fairly constant and high in the lower reaches, reducing rapidly in the middle estuary, and again constant but low in the upper estuary. Absolute values depend upon the freshwater input from the river

systems. An example of the salinity close to the surface along the Humber is given in Figure 1.11b, taken from Gameson (1982).

1.9 PARTICULATES

Estuaries represent one of the most important transport paths for the transfer of sediment between land and ocean systems. Uncles et al. (2001) estimate that most of the 20×10^9 T of global suspended sediment transferred each year passes through estuarine environments generating concentrations in excess of 1,000 mg dm^{-3}. Estuaries exhibit strong horizontal gradients in SPM with an estuarine turbidity maximum (ETM) which is advected by freshwater and tidal processes. The result is, therefore, complex, yet observations in many estuaries attest to some underlying principles. The most recent research results are discussed in Chapter 11. Figure 1.11c shows the variation in concentration of near surface suspended particulate matter along the Humber taken from Uncles et al. (1998).

1.10 CLASSIFICATION OF ESTUARIES

There are, within the extensive estuarine literature, a number of classifications that make reference to the fundamental differences between estuaries which are dominated by the sea and those which are dominated by fluvial processes. For example, Dyer (1986) presents a description and classification of estuaries based upon the vertical structure and the mixing of the saltwater and the freshwater and recognizes salt wedge, partially mixed, and well-mixed estuarine environments. Alternatively, Woodroffe (2003)

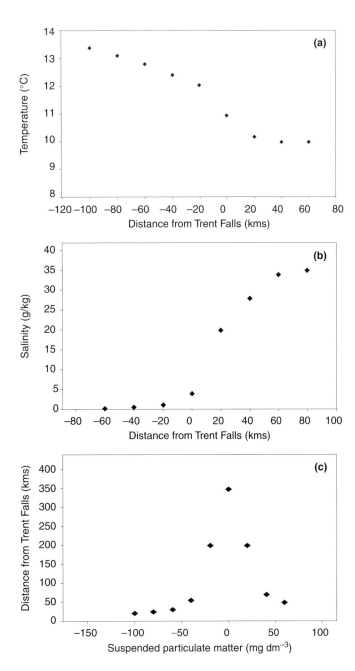

FIGURE 1.11 Water Properties in the Humber Estuary (a) temperature, (b) salinity, and (c) suspended particulate matter (after Gameson, 1982(a and c) and Mallowney, 1982(b)).

classifies estuaries in accordance with their sedimentation patterns as ranging from drowned river valleys to tide-dominated and wave-influenced estuarine environments.

Salt wedge estuaries

Highly stratified estuaries with a saline lower layer originating from the sea are known

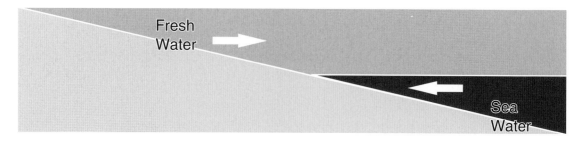

FIGURE 1.12   Diagrammatic representation of circulation in a salt wedge estuary (after Dyer, 1986).

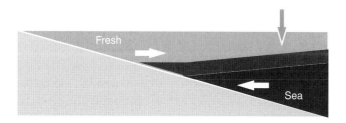

FIGURE 1.13   Diagrammatic representation of circulation in a partially mixed estuary (after Dyer, 1986).

as **salt wedge estuaries** (Figure 1.12). Salt wedge estuaries are maintained by low tidal velocities and hence weak vertical mixing and evidence a freshwater upper layer originating from the river discharge. Examples of salt wedge estuaries include the Mississippi in America and the Rhone in France. The position of the salt wedge will vary with river flow. A typical example of a salt wedge estuary is the South West Pass of the Mississippi. When the flow is low the salt wedge extends more than 100 miles inland, but with high river discharge the salt wedge only reaches a mile or so above the mouth.

Partially mixed estuaries

In **partially mixed estuaries**, the water mass moves seaward and landward with tide-generated mixing between the salt water and the fresh water (Figure 1.13). Examples of partially mixed estuaries include

the Rotterdam waterway in The Netherlands and the Potamac in America.

Well-mixed estuaries

Well-mixed estuaries exhibit complete mixing of saltwater and freshwater (Figure 1.14) resulting in a constant fluid density over the depth and a varying density in longitudinal direction from that of seawater ($1,025 \, \text{kg m}^{-3}$) to that of freshwater ($1,000 \, \text{kg m}^{-3}$). Examples of well-mixed estuaries include the Severn in England and the Gironde in France, the Western Scheldt in The Netherlands, and the Mersey in England.

The type of estuary that develops depends upon the interaction between the influences of the tidal processes, nearshore wave processes, and fluvial inputs from the river system. The results are complex and there exists a wide variety of estuarine forms around the coasts of the world's oceans. However, there are some generalizations

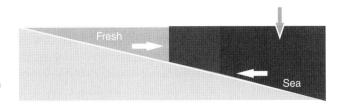

FIGURE 1.14 Diagrammatic representation of well-mixed estuary circulation (after Dyer, 1986).

that can be drawn. Estuaries can be separated into three main groups:

1 **Drowned valleys**, particularly those in which the prior bedrock configuration exerts a primary control on morphology, such as rias and fjords;
2 **Tide-dominated estuaries** associated with large tidal ranges;
3 **Barrier estuaries**, in which the longshore transport of coarse sediment has a particular influence on the morphology of the estuary's entrance.

Drowned valley estuaries: fjords and rias

The Nordic term fjord refers to glaciated troughs with mountainous shores that occur poleward of 45° N and S particularly in western Canada, Norway, southern Chile, Greenland, Labrador, Iceland, Antarctica, southwestern New Zealand, and in Scotland where they are called "lochs." The entrance to the fjord often exhibits a sill and the enclosed basin is often deep, plunging for example more than 2,000 m deep in Antarctica (Syvitski et al., 1987). Circulation in fjords is often stratified, and the deep basin can be stagnant and anoxic because the sill hinders exchange. A four-stage model is usually advanced to account for the evolution of fjords:

1 Ice-filled during the glacial event as with many fjords in Greenland;
2 Becomes filled with seawater and floating ice as on Baffin Island in Canada;

3 Ice only remains on shore as in many fjords in Alaska and Norway;
4 Ice free leaving U-shaped basins, such as the sea lochs of Scotland and Milford Sound in New Zealand.

Rias, however, are V-shaped drowned valleys where the bedrock topography plays a prominent role in estuarine morphology, as opposed to the alluvial systems described below. The term "ria" is Spanish and is used for the high relief estuaries of the Iberian Peninsula (e.g. the Ria de Muros y Noya (Somoza and Rey, 1991) and the Sado in Portugal (Psuty and Moreira, 2000)). Rias are also identified in Brittany (where the term "aber" is used), in southwest England (the Dart (Thain et al., 2004) and Tamar (Uncles et al., 1994)), and along the coast of New South Wales in Australia and eastern North America (Woodroffe, 2003).

Tide-dominated estuaries

Estuaries that occur in a macrotidal setting are dominated by tidal currents and, in accretionary environments, are typically funnel shaped with wide entrances tapering upstream (Chappell and Woodroffe, 1994; Wells, 1995). These estuaries are well mixed as a result of strong tidal currents which suspend and transport sediment. Dalrymple et al., (1992) recognize three characteristic zones:

1 Upstream river-dominated zone, usually relatively straight with net seawards movement of sediment by fluvial processes;

2 Central zone exhibiting sediment convergence influenced by both fluvial and marine processes and with a highly sinuous channel;

3 Marine-dominated seaward zone which is relatively straight and funnel shaped and sediment moves in a net landward direction.

There are tide-dominated estuaries associated with the macrotidal regions of northern Australia (e.g., Darwin Harbor (Semeniuk, 1985), the Orde, South Alligator, and Daly Rivers (Chappell, 1993)), the Humber on the eastern coast of England (Hardisty et al., 1998), and Cobequid in the Bay of Fundy, Canada (Dalrymple et al., 1990). A three-stage evolution of tide-dominated estuaries is often invoked consisting of (Woodroffe, 2003):

1 Transgressive phase as the sea-level rises;
2 "Big Swamp" phase at the time of sea-level stabilization;
3 Stillstand phase including recent floodplain deposition.

Barrier estuaries

On wave-dominated coasts, barrier (or barbuilt) estuaries are generally enclosed or partially enclosed behind wave-built sand barriers (Woodroffe, 2003). The inlet that connects to the sea can intermittently close, and there is gradation into coastal lagoons that may be permanently closed or intermittently open to the sea. Barrier estuaries are, typically, accreting due to a combination of three depositional processes (Roy, 1984):

1 Input of sand and finer sediment from seaward;
2 Extension of fluvial deltas of sand and mud at the mouths of rivers and streams within the estuary; and
3 Vertical accretion of mud throughout the central estuary.

Roy (1984) and Roy et al. (1994) describe examples of barrier estuaries in various stages of the accretionary cycle.

2

MONITORING ESTUARINE SYSTEMS

Contents

2.1 INTRODUCTION

This book considers estuaries in terms of the shape of the basin and the processes associated with the vertical rise and fall of the tide, the freshwater and tidal currents, the advection and diffusion of salt and thermal energy, and finally the transport of particulates. Each of these subjects is dealt with in a separate section of the book. The development of monitoring techniques for each parameter is described in this chapter.

2.2 BATHYMETRIC SURVEYING

Bathymetric surveying demands two practical requirements which, until relatively recently, have been difficult to achieve. First, the surveyor must locate his or her position on the sea surface and, second, he or she must determine the depth of water at that location and correct the depth onto one of the different data detailed below. There is a wide variety of methods for estuarine position fixing which vary from simple optical sextant fixes to electronic systems linked to orbiting satellites.

Manual position fixing

Close to the coast the observer may use simple compass bearings or a vertical sextant to determine the distance from a coastal landmark. With less accuracy, celestial observations with a sextant can be combined with careful timing and tables of astronomical constants to determine the ship's position at sea. These optical systems demand good visibility and a relatively calm sea; only in the hands of an experienced seaman will reliable results be obtained from a heaving deck.

Electronic position fixing

Older electronic systems are based upon the comparison of signals from shore-based radio stations. The Decca Navigator system is the primary aid of this type for ships operating in the seas around the British Isles. It consists of chains of radio stations which transmit low-frequency radio signals. The signals intersect in a known hyperbolic pattern and a receiver on the ship displays the results on three Decometers as sets of letters and numbers. Each letter and number identifies a line on the chart and the intersection of lines from two of the Decometer readings allows a position to be plotted.

More recently, microwave, range and bearing, and combined systems have been introduced in coastal locations and effectively replace the line of site optical systems by providing electronic distance and direction information with respect to shore-based transmitters. Most recently, satellite navigation systems have been introduced which measure the distance of one or more orbiting or geostationary satellites. The systems also gather information on the height and position of the satellites, and can then compute the ship's position with remarkable accuracy. It is likely that satellite navigation systems will supersede all of the other position-fixing methods in the future.

Manual depth determination

The depth of water at any given location was, until the middle of the last century, determined by the traditional lead and line soundings. In shallow water this proved sufficiently accurate for most purposes and tallow wax was often inserted into the base of the lead to recover a sample of seabed sediment. However, the technique is laborious and can, in deep water, be extremely time consuming and prone to inaccuracies. Linklater (1972), writing about the famous voyage of the *Challenger* in 1872, notes that a single deep sea depth measurement often consumed a whole day's work and that the errors due to a nonvertical line or wire could be considerable.

Acoustic depth determination

The advent of underwater acoustics in the early twentieth century led to the development of echo-sounders which measure the two-way travel time of a pulse of sound from the vessel to the seabed and back, and convert this to water depth. Mechanical echo-sounders usually employed a belt-driven or radial-arm-driven electrical stylus passing across a strip of specially prepared paper to provide a continuous profile of the seabed. A higher stylus speed, and hence faster pulse repetition rate, was used with surveying systems than in conventional navigational echo-sounders, in order to provide maximum resolution of the seabed and a comparatively large vertical scale. Mechanical instruments have largely been replaced by electronic and digital devices. Typically, transmission frequencies lie in the band 30–210 kHz. Lower frequencies are used in deep water because of the attenuation of the higher frequencies.

2.3 TIDE GAUGES

Estuarine tides are observed as the regular rise and fall of sea-level due to the gravitational attraction of the Moon and the Sun on the waters of the world's oceans. Most estuaries experience approximately two tidal cycles in each twenty-four-hour period and the range of the tide increases from small values in the deep ocean to many meters in certain coastal and estuarine locations. For example, the Minas Basin at the upper end of the Bay of Fundy and Lac Aux Feuilles in Hudson Bay experience ranges in excess of 16 m and the Severn Estuary in England has only slightly smaller ranges.

The monitoring of tidal elevations in an estuary aims to measure the vertical distance between the water surface and a fixed datum. Although this may appear to be a trivial exercise in a laboratory, the situation in an estuary is complicated by, for example, the need to smooth the readings to remove the effects of short-period wind waves, the requirement to protect the system against bio-fouling, and even the rampages of site vandalism. The following section is based in part upon Hardisty (1990a) and Pugh (1987 and 2004). A more detailed description of the techniques involved in choosing and operating a tide gauge system is given in UNESCO's manual, Intergovernmental Oceanographic Commission (1985).

Pugh (2004) relates that, whilst Isaac Newton provided the theoretical explanation for tidal phenomena, it was one of his contemporaries, Edmond Halley, who first made systematic measurements of tides in the English Channel in the late seventeenth century. Francis Beaufort (of wind scale fame) analyzed 30-year records of high and low water at London Docks and published the results in the *British Almanac* for 1830. Lubbock then set out to record "the heights and times of high water over a long period to calculate the different variations to be detected in the tides, the semi-diurnal rise and fall, the twice-monthly springs and neaps, the alterations caused by the variation in the moon's orbit, and to relate them to the astronomical forces which had caused them" (Deacon, 1971). This initiative was taken up by the Royal Society which appealed to the Admiralty who agreed to install automatic tide gauges in the dockyards at Sheerness, Portsmouth, and Plymouth, and two sites in London in 1831.

Lubbock's Cambridge tutor, William Whewell, widened the field "to discover the general pattern of tides throughout the world." Beaufort again assisted by organizing simultaneous measurements around Britain and Ireland in June 1834, extending to "both sides of the Atlantic, at twenty eight or more places between Florida and Nova Scotia and from the Straits of Dover to North Cape" in 1835. James Clark Ross is recorded as obtaining tidal data whilst trapped in the ice during the Arctic winter of 1848–9.

Tide poles

Tide Poles or staffs (Figure 2.1) are cheap, easy to install, and perfectly adequate for a short-term survey. The pole is graduated and permanent installations are often found around harbors to assist shipping and navigation. It is necessary to average the readings over a short time interval to remove the effects of wave action. However the tedium involved and the errors that result if readings are required over a longer period of time makes tide poles unsuitable for long-term, high frequency data.

FIGURE 2.1   Tide pole at Anchorage, Alaska (DIR 2.1).

The Palmer–Moray gauge

The vast majority of permanent gauges installed during the nineteenth and twentieth centuries consisted of an automatic chart recorder: the recording pen is driven by a float which moves vertically in a well, connected to the sea through a relatively small hole or narrow pipe. The limited connection damps the external, short-period wave oscillations.

The basic idea was described for the Royal Society of London by Moray in the seventeenth century. He proposed that a long narrow float should be mounted vertically in a well, and that the level to which the float top had risen be read at intervals. The first, self-recording gauge, designed by Palmer, began operating at Sheerness in the Thames estuary. Figure 2.2 shows the components of such a Palmer–Moray system. The well is attached to a vertical structure which extends below the lowest level to be measured and has one or more small connections

to the open sea. Copper is often used to prevent fouling of the narrow connections by marine growth and a layer of kerosene may be poured over the seawater in the well where icing is a winter problem.

Vertical float motion is transmitted by a wire or tape to a pulley and counterweight system. Rotation of this pulley drives a series of gears which reduce the motion and drive a pen across the face of a chart mounted on a circular drum. Usually the drum rotates once in 24 h, typically at a chart speed of $0.02\,\mathrm{m\,h^{-1}}$; the reduced level scale may be typically 0.03 m of chart for each meter of sea-level. Alternative forms of recording included both punched paper tape and magnetic tape.

Although stilling-well systems are robust and relatively simple to operate, they have a number of disadvantages. They are expensive and difficult to install, requiring a vertical structure for mounting over deep water. Accuracies are limited to about 0.02 m for levels and 2 min in time because of the width of the chart trace. Charts can change their dimensions as the humidity changes. Reading of charts over long periods is a tedious procedure and prone to errors. Palmer–Moray gauges are being replaced by the electrical or electronic devices detailed in the following sections.

Pressure gauges

The hydrostatic pressure at a fixed point below the water surface is related to the overlying water depth:

$$P = P_a + \rho g h \qquad \text{Eq. 2.1}$$

where P is the hydrostatic pressure, P_a is the atmospheric pressure, ρ is the density of seawater (about $1040\,\mathrm{kg\,m^{-3}}$), g is the gravitational acceleration ($9.81\,\mathrm{m\,s^{-2}}$), and h is the water depth.

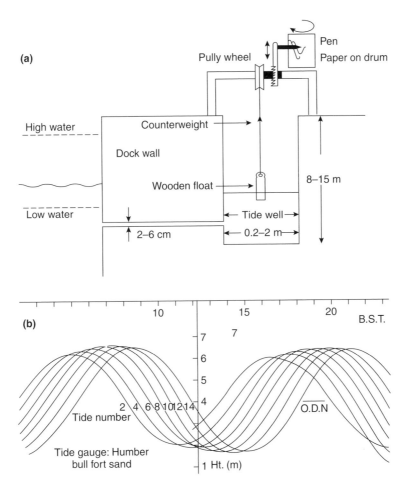

FIGURE 2.2 A Palmer–Moray stilling-well system (a) and typical tidal record (b) from the Bull Fort gauge in the Humber Estuary (after Hardisty, 1990a).

Devices which measure this pressure and thus the water depth are called pressure gauges or pressure transducers. Direct reading and recording systems are available from a number of commercial organizations (Figure 2.3).

In many systems, the pressure transducer is kept dry in the bubbler gauge. Compressed air is fed into an upper chamber at a pressure which bubbles out at the submerged end of the sensing pipe. The pressure throughout the system will thus be equivalent to the pressure at the orifice, and the system then determines this pressure and hence the depth of water.

Acoustic tide gauges

The time taken for a pulse of sound to travel from the source to a reflecting surface and back again is a measure of the distance from the source to the reflector. The travel time is given by

$$t = \frac{h}{C_0} \qquad \text{Eq. 2.2}$$

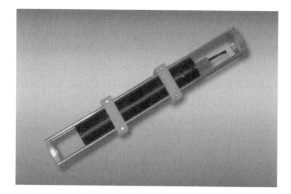

FIGURE 2.3 Deep water pressure transducer gauge (DIR 2.2).

where t is the travel time (s), h is the path length (m), and C_o is the velocity of sound (m s^{-1}).

This principle is utilized in the acoustic tide gauges which are being installed in many locations as shown in Figure 2.4. Such devices send an acoustic pulse down a transmitting tube. The tube has a fixed reflector above the water surface which is used automatically to calibrate the device and the travel time for the water surface echo is used to measure the location of the water surface and hence to monitor the tidal elevations.

Online tidal information

The development of the Internet has enabled an increasing number of online tidal data sources to be accessed. There are long-term, archival datasets such as the British Oceanographic Data Centre, and operational sites showing real-time or near-real-time data, often in comparison with predictions as at the CO-OPS sites. A selection is described below.

British Oceanographic Data Centre: DIR 2.4 – The UK National Tide Gauge Network, run by the Tide Gauge Inspectorate,

FIGURE 2.4 Acoustic tide gauge installed at Settlement Point, Bahamas (DIR 2.3).

records tidal elevations at 44 locations around the UK coast

Tide Online: DIR 2.5 – NOAA's National Ocean Service (NOS) operate a large number of online tide gauges around the North American coastline and across the Pacific

ACCLAIM: DIR 2.6 – The ACCLAIM program in the South Atlantic and southern oceans consists of measurements from coastal tide gauges and bottom pressure stations, together with an ongoing research program in satellite altimetry.

2.4 CURRENT METERING

This section considers the manual, mechanical, electronic, and automated techniques that have been developed to measure the speed and direction of water flows in estuarine environments. The problem is not trivial: estuarine currents, typically, increase from zero at slack water to maxima which can be in excess of $2-3\,\mathrm{m\,s}^{-1}$ at mid tide before decelerating, reversing, and then accelerating to achieve similar values in the opposite direction. Estuarine current metering, unlike work in streams and rivers, must therefore not only be capable of resolving a wide range of speeds, but must also encompass vector as opposed to simple scalar data.

The measurement of water flow and currents in estuarine environments combines the interest of oceanographers and hydrographers with branches of the environmental and engineering sciences. Background information may be found in such text books as Neumann and Pierson (1966). The techniques separate into Lagrangian with a sensor mounted at a fixed point and data describe time series of the flow vectors at that point, and Eulerian which track the path of a water particle.

The measurement of tidal and freshwater flow in estuaries probably dates back almost as far as the measurement of the tidal water depth itself. Early charts, for example, included descriptions of flows based upon the nautical term of knots. One knot is equivalent to a flow speed of $0.51\,\mathrm{m\,s}^{-1}$. Mechanical devices embodying some form of rotating element which are used for water velocity measurement are called *current meters* and *The Proceedings of the Royal Society of London* record descriptions of such devices from Newton's time.

Mechanical flow sensors

Rotating cup current meters work because the drag on the hollow hemisphere or cone is greater when its open side faces the flow, and so there is a net torque on the assembly that causes the cups to rotate. The magnitude of the fluid velocity

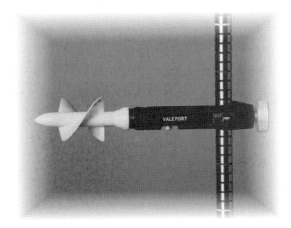

FIGURE 2.5   *Valeport Braystoke impellor flow meter (DIR 2.7).*

determines the speed of rotation, which is usually indicated by the number of once-per-revolution counts made in a known time interval. For higher velocities, an impellor is used having its axis parallel to the fluid flow (Figure 2.5). The impellor current meter is sensitive to the direction of flow particularly if the propeller is surrounded by a shielding cylinder.

Acoustic flow sensors

There are two, quite different, acoustic techniques for the determination of water flow speeds and these are known as acoustic travel time and acoustic Döppler devices. In the former, the travel time of an acoustic pulse between a source and a receiver depends upon three factors:

1 The distance between the source and the receiver;
2 The speed or celerity of the sound (about $1410\,\mathrm{m\,s}^{-1}$ in seawater); and
3 The velocity of the flow: if the flow travels from the source to the receiver, then the travel time is reduced and *vice versa*.

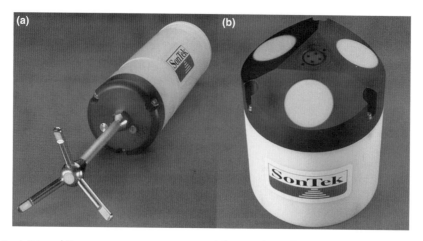

FIGURE 2.6 (a) Travel Time acoustic current meter and (b) Döppler shift acoustic current meter (DIR 2.8).

Acoustic travel time current meters (Figure 2.6a) measure the travel time in, for example, three dimensions to compute the current speed vector. Acoustic Döppler current meters (Figure 2.6b), emit a sound pulse which is detected after it has been reflected back from particles in the flow. The reflected sound will evidence a Döppler frequency shift which depends upon the speed and direction of the flow. The instruments measure the Döppler shift and compute the flow velocity.

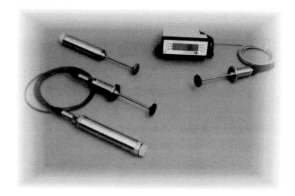

FIGURE 2.7 A range of electromagnetic current meters (DIR 2.9).

Electromagnetic flow sensors

It has long been appreciated that movement of a conducting medium, such as salt water, around a primary electrical coil induces an electromotive force in a secondary coil. Electromagnetic current meters and flow sensors utilize this effect by measuring the induced current in a secondary coil encapsulated in a sensing head (Figure 2.7) to compute the current velocity.

Eulerian techniques

Among special flow measurement techniques used in rivers, estuaries, and the

open sea is the *salt-velocity method*, in which a quantity of a concentrated solution of salt is injected into the water. At each of two downstream cross-sections, a pair of electrodes is inserted, the two pairs being connected in parallel. When the salt-enriched water passes between a pair of electrodes its higher conductivity causes the electric current to rise briefly to a peak. The mean velocity of the stream is determined from the distance between the two electrode assemblies and the time interval between the appearances of the two current peaks. Rather elaborate apparatus is needed

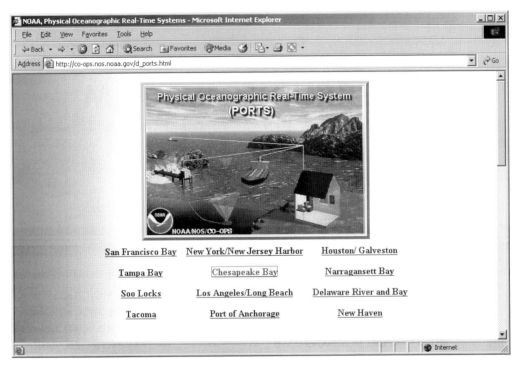

FIGURE 2.8 The North American PORTS system (DIR 2.10).

for the automatic recording of data. More pragmatically, floats are frequently tracked to determine mean current drift directions.

Online current data

One of the most interesting sources of online estuarine flow data is the Physical Oceanographic Real Time (PORTS) system operated in North America (Figure 2.8).

2.5 THERMOMETRY

The salinity and temperature of water have been monitored throughout recorded history using manual techniques. The second half of the twentieth century witnessed the introduction of the automated and electronic systems described here. This section

is based, in part, on Hardisty (1990a) and Ridout (2004).

Manual temperature determinations

Most substances expand when heated, and this property has been used for centuries in the manufacture of thermometers. Most commonly, "mercury in glass" thermometers are used. There are problems, however, associated with the use of such thermometers for measuring the temperature of estuarine waters. The thermometer is usually lowered into the water to determine the temperature at the required depth. However, the reading changes during recovery as the instrument is hauled to the surface through water which may have a different temperature and in the air. The problem was resolved with the introduction of

(a)

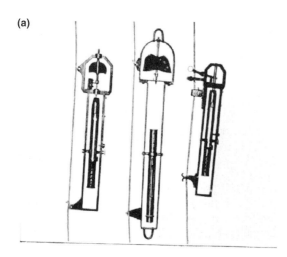

(b)

FIGURE 2.9 Traditional oceanographic reversing thermometer (DIR 2.11).

the oceanographic reversing thermometer as shown in Figure 2.9.

Electronic thermometers

Water temperature is measured electronically with thermistors and thermocouples which sense temperature-dependent changes in the electrical properties of conductors (Figure 2.10).

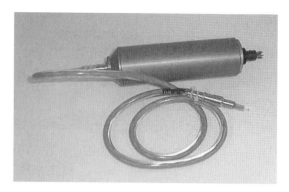

FIGURE 2.10 Seabird Electronics oceanographic temperature sensor for use in estuarine waters (DIR 2.12).

2.6 Estuarine salinity determinations

Early work on measuring the saltiness of the sea involved taking a known weight of seawater, evaporating the water, and weighing the residue (Boyle, 1693; Birch, 1965). Refinements to the analysis included solvent extraction (Lavoisier, 1772) and precipitation (Bergman, 1784) using an Ocean Salinometer (Figure 2.11). Forchhammer (1865) introduced the term salinity and the concept of measuring one parameter, chloride, from which the salinity could be calculated. This work was supported further by Dittmar (1884) who analyzed samples from the Challenger Expedition to establish the theory of "Constant Composition of Seawater". These methods remained dominant until the 1950s when conductivity was introduced as a practical means to measure salinity.

Chemical salinity determinations

Since chloride ions are one of the major constituents of estuarine water, and the relative proportions are constant, it is only necessary to determine the chlorinity (Cl‰) of a sample. The oldest, and

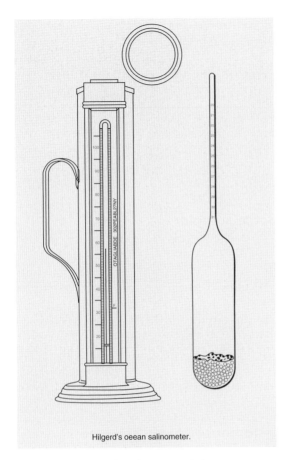

FIGURE 2.11 Hilgard's Ocean Salinometer
(c. 1880) (DIR 2.13).

FIGURE 2.12 University of Miami salinometer on board the Royal Caribbean cruise ship *Explorer of the Sea* (DIR 2.14).

of sea water when all of the carbonate has been converted to oxide, all the bromine and iodine replaced by chlorine and all the organic material oxidised".

still often applied, method is based upon Mohr's Cl$^-$ titration, where the chlorine content is determined with silver nitrate using a potassium chromate solution as the indicator. The salinity is then given by the internationally agreed empirical relationship:

$$S‰ = 0.030 + 1.8050Cl‰ \qquad \text{Eq. 2.3}$$

Further work by Knudsen et al. (1902) resulted in a new definition which stated that salinity was "The total amount of solid material in grams contained in one kilogram

Electronic salinity sensors

Salinity is now determined on a routine basis with electrical instruments which measure the conductivity of the water sample and are calibrated to directly calculate the value of either the chlorinity or salinity (Figure 2.12). Various salinometers were developed during the 1950s and 1960s (e.g. Hamon, 1955) and new relationships were produced for conductivity and salinity (Millero et al., 1977; Poisson et al., 1978). These gave rise to the widely accepted "industry standard" salinometers (Autosal and Portasal) which are produced by Guildline Instruments Ltd. and achieve a quoted accuracy of ±0.002 in practical salinity units.

2.7   ESTUARINE PARTICULATES

The principles of field monitoring of suspended particulate matter are usually contained in the research papers referenced below, but introductory discussions may be found in such as Sternberg (1986). One of the better discussions of the techniques and errors in particulate monitoring is given by Van Rijn (1993).

Particulate determinations were based upon the range of techniques for obtaining a water sample and removing the solids. The techniques which measure the effect of the particulates on acoustic or optical propagation through an *in situ* water sample were introduced more recently.

Although the determination of the concentration of SPM in estuarine waters is a relatively recent endeavor, the study of the "muddiness" or "turbidity" of water has earlier origins. Since the early days of oceanographic endeavor in the mid-nineteenth century, such work has utilized the Secchi disk as shown in Figure 2.13. The disk consists of an 8″ (23 cm) diameter black and white plate which is lowered into the water until no longer visible and then raised until again visible. The Secchi depth (of extinction and usually in feet) is then recorded as an indication of the turbidity of the water.

The Secchi depth was roughly correlated with the Jackson turbidity units (JTU's) which were introduced at the end of the nineteenth century through the Jackson Tube. This device is a long glass tube suspended over a lit candle. A sample of water was slowly poured into the tube until the candle flame, as viewed from above, could no longer be seen and the depth of water was taken as the turbidity in JTUs. Nowadays, turbidity is measured in nephelometer turbidity units (NTU) with the instruments described below.

Gravimetric methods

There are a range of techniques for the determination of the concentration of particulates in estuarine systems which involve sampling the water, usually at a point and for a limited time interval, and then removing (through filtration or drying) and weighing the solids to determine the concentration as illustrated in Figure 2.14.

Bottle sampler

Most simply, a stoppered bottle is lowered to the required depth within the water column and a line or electric relay is used to remove the stopper. Water and particulates flow into the bottle which is then returned to the laboratory for analysis (Delft Hydraulics, 1980). There are problems associated with the natural fluctuations in the actual concentration in the field for short filling times.

Instantaneous trap sampler

The instantaneous trap sampler consists of a horizontal cylinder equipped with end valves which can be closed suddenly (by a messenger system) to trap a sample instantaneously. The water is allowed to flow freely through the open horizontal cylinder while the sampler is lowered to the desired point. Again the disadvantages are that the sampler is incapable of detecting natural variations in the local concentration.

Time-integrating trap sampler

This consists of a vertically hung funnel or cylinder with a closed bottom. The device traps particulates in its enclosure and the open top is closed by a door or latch for recovery. The efficiency of such systems

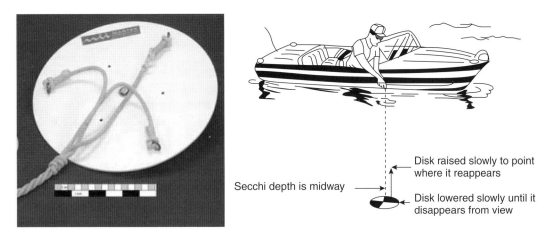

FIGURE 2.13   The principles of the Secchi disk for measuring *in situ* turbidity.

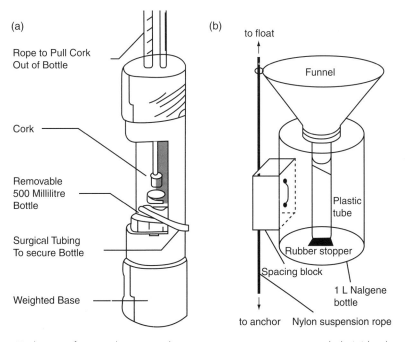

FIGURE 2.14   Techniques for sampling particulate matter in estuarine waters include (a) bottle sampler and (b) time-integrating trap sampler.

decreases from 100% in still water as the current speed increases and sediment is swept out before recovery (Bloesch and Burns, 1980). A development is the streamer trap (Figure 2.14) wherein the water flows into the device and the particulates are filtered out as the water exits and are described by Krause (1987), and the bottle sampler (Figure 2.14a) where internal tubes arrange for the sediment to settle out before the water is released downstream.

Pump sampling

Finally, samples of the water can be obtained and returned to the shore or to a survey boat using carefully arranged pipes and pumps. Particulates in the sample are either settled out or removed with a filter to determine the concentration

Optical methods

Optical and acoustic methods for the determination of particulate concentrations are based on the effect of the presence of the particles on the reflection or propagation of a beam. Optical methods separate into transmissometers, nephelometers, and back scatter devices as illustrated in Figure 2.15.

Transmissometers

These consist of a light source, a straight light path, and a light detector. The amount of light which is transmitted from the source to the receiver depends inversely on the concentration of particulates (Figure 2.16):

$$I_t = k_1 \exp(-k_2 C) \qquad \text{Eq. 2.4}$$

where I_t is the intensity of the light received at the detector, C is the concentration of particulates, and k_1 and k_2 are calibration coefficients for the device and shape and composition of the sediment.

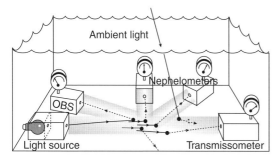

FIGURE 2.15 The principles of optical methods including back scatter and transmissometers.

Nephelometers

These consist again of a light source and a detector that are mounted at about 90° to the illuminated beam. The fraction of the light that is scattered into the detector increases with the concentration of particulates and is given by

$$I_s = k_3 C \exp(-k_2 C) \qquad \text{Eq. 2.5}$$

where I_s is the light scattered to the detector, C is again the concentration of the particulates, and k_2 and k_3 are calibration coefficients which depend again upon the shape and optical properties of the particulates.

Back scatter devices

These consist of a light source and a detector that is mounted adjacent to the source (Figure 2.17) at 180° to the illuminated beam. The fraction of the light that is scattered into the detector increases with the concentration of particulates and is also given by Eq. 2.5.

Acoustic methods

Single frequency acoustic backscatter measurements of suspended sediments in the ocean were introduced in the 1980s (Young et al., 1982; Hay, 1983; Hess and Bedford, 1985; Lynch 1985) using frequencies in the range 1–4 MHz. More recently, Thorne et al. (1991) have developed a downward directed acoustic concentration sensor working on backscattering principles known as the Acoustic Profiling System. A short pulse (10 μs) of high frequency (1–5 MHz) is emitted from a transducer and some of the acoustic energy is scattered back from particulates in the water column to the transducer. The magnitude of the backscattered

FIGURE 2.16 Partech transmissometer for suspended solid monitoring and optical back scatter (OBS-3) device (DIR 2.15).

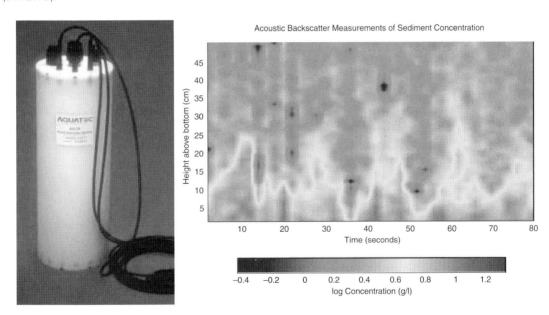

FIGURE 2.17 Acoustic backscatter device and example of suspended particulate concentration profilers obtained on the Californian continental shelf (DIR 2.16).

signal is related to the concentration, particle size, and the time delay between transmission and reception. Range gating the detected signal permits the concentration profile to be estimated (Vincent et al., 1991).

Remote sensing

Measuring SPM remotely relies on the fact that the reflectance of visible light increases with the amount of suspended sediment in the surface waters. The satellite sensor used is the Advanced Very High Resolution Radiometer (AVHRR), which has been in operation since the early eighties. Initially a relationship between reflectance and the suspended sediment was found using field data collected on research cruises. An optical instrument was used to measure reflectance at sea which mimicked the measurements made by the AVHRR. The satellite images can be processed with this relationship to make maps of sediment concentrations (White, 1999).

Online particulate data

There are nowadays long-term, archival data and operational sites showing real-time or near-real-time data for suspended sediment. Two examples are as follows:

Newfoundland: DIR 2.17 – Near-real-time data from the Canadian Humber River.

North Sea: DIR 2.18 – West Gabbard buoy in the southern North Sea is a Smart Buoy operated by CEFAS.

Part II

THE BATHYMETRY OF ESTUARIES

3

ESTUARINE BATHYMETRY

Contents

3.1 INTRODUCTION

The three-dimensional shape of an estuary, and the manner in which that shape changes over time, are known as the bathymetry and bathymetric evolution, respectively. Bathymetric surveys, employing modern acoustic and position fixing techniques (Chapter 2) have revealed that the accretionary, tide-dominated estuaries described in Chapter 1 exhibit certain simple width–depth relationships within a complex bathymetry. These relationships are described here and incorporated into the estuarine model in the following chapter.

3.2 A BRIEF HISTORY OF HYDROGRAPHY

Bathymetric surveying has long been the provenance of the world's navies and is one element of the British Admiralty's Hydrographic Office (DIR 3.1). The first hydrographer to the British Admiralty, Alexander Dalrymple FRS, was appointed by order of King George III in 1795. He set to work reviewing the "difficulties and dangers to his Majesty's fleet in the navigation of ships," but the first Admiralty Chart (of Quiberon Bay in Brittany) did not appear until 1800. Under the second hydrographer, Captain Thomas Hurd, who served from 1808 to

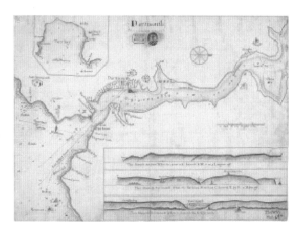

FIGURE 3.1 Early Admiralty chart of the Dart Estuary, Devon in England (DIR 3.1).

1823, permission was given to sell charts to the public, and he oversaw the production of volumes of sailing directions and the first chart catalogue (Figure 3.1).

The fourth hydrographer, Rear Admiral Sir Francis Beaufort KCB, FRS, was appointed in 1829 and worked tirelessly until 1855, organizing surveys all over the world to expand and improve chart coverage. Among his many accomplishments were the Beaufort Scale of wind strength, the introduction of official Tide Tables in 1833 (see also Chapter 5), and Notices to Mariners in 1834.

The skills of surveying and chart production were consolidated and world coverage increased during the second half of the nineteenth century, whilst the demands of the two World Wars gave impetus to technical innovations in instruments and techniques during the twentieth century. The development of the echo-sounder in the 1930s and side-scan sonar in the 1960s brought huge advances in the bathymetric charting of the sea bed.

3.3 CHARTED DEPTHS

Estuarine bathymetry is usually presented for navigational purposes on hydrographic charts with reference to a Chart Datum (CD). CD approximates to the lowest astronomical tide which is defined as the lowest water level that can be predicted to occur under average meteorological conditions. Other levels are often specified with reference to this datum:

Highest astronomical tide	**HAT**	Maximum tidal water level
High water springs	**HWS**	Maximum mean spring tide
High water neaps	**HWN**	Maximum mean neap tide
Low water neaps	**LWN**	Minimum mean neap tide
Low water springs	**LWS**	Minimum mean spring tide
Lowest astronomical tide	**LAT**	Minimum tidal water level

These levels are often compared with the local topographic datum that, in the UK, is known as Ordnance Datum Newlyn (ODN) as described in Pugh (2004). ODN is the mean sea-level at Newlyn measured between 1915 and 1921, and remains fixed even though present mean sea-level at Newlyn has risen by about 0.20 m since that determination.

The definition of the CD in terms of tidal behavior implies that it is not A horizontal surface, but changes with the tidal range around a coast or within an estuary. For example, in the Humber, the relationships are as follows:

	HWS	HWN	LWN	LWS	CD
BULL FORT ODN	6.9	5.5	2.7	1.1	3.9 m below
IMMINGHAM ODN	7.3	5.8	2.6	0.9	3.9 m below
HULL ODN	7.3	5.3	2.4	0.8	3.9 m below
HESSLE ODN	7.3	5.3	2.2	0.4	3.4 m below

TABLE 3.1 Estuarine coefficients (from Prandle (1986)).

	Length (km)	n	m
Fraser	135	−0.7	0.7
Rotterdam	99	0	0
Hudson	248	0.7	0.4
Potomac	184	1.0	0.4
Delaware	214	2.1	0.3
Miramichi	55	2.7	0
Bay of Fundy	635	1.5	1.0
Thames	95	2.3	0.7
Bristol Channel	623	1.7	1.2
St Lawrence	418	1.5	1.9

estuary (m), and λ is a horizontal dimension taken as the length of the estuary (m).

Values for the two coefficients, m and n, are given by Prandle (1986) for major estuaries and reproduced here in Table 3.1. The results are plotted in Figures 3.2 and 3.3, showing the classical, flaring estuarine width and exponentially increasing depth.

3.4 WIDTH AND DEPTH AS FUNCTIONS OF DISTANCE

Whilst working on generalized, analytical theories for estuarine dynamics, Prandle (1986) argued that the bathymetry of many estuaries may be approximated by the functions

$$W_x = W_L \left(\frac{x}{\lambda}\right)^n \qquad \text{Eq. 3.1}$$

and

$$D_x = D_L \left(\frac{x}{\lambda}\right)^m \qquad \text{Eq. 3.2}$$

where W_x is the estuary width (m), W_L is the estuary width at the mouth, D_x is the estuary depth (m), D_L is the estuary depth at the mouth, x is measured from the mouth of the

3.5 WIDTH AND DEPTH AS EXPONENTIAL FUNCTIONS OF DISTANCE

Dyer (1986) notes that many estuaries show an exponentially varying width, depth, and cross-sectional area with distance from their heads. In a similar manner, Prandle (1986) replaces Eq. 3.1 and Eq. 3.2 with

$$W_x = W_0 e^{-nx} \qquad \text{Eq. 3.3}$$

$$D_x = D_0 e^{-mx} \qquad \text{Eq. 3.4}$$

These are analogous to Wright et al. (1973) showing that the width can be represented by

$$W_x = W_0 e^{-a(x/L)} \qquad \text{Eq. 3.5}$$

where W_x is the estuary width (m), L is the estuary length (m), x is the distance

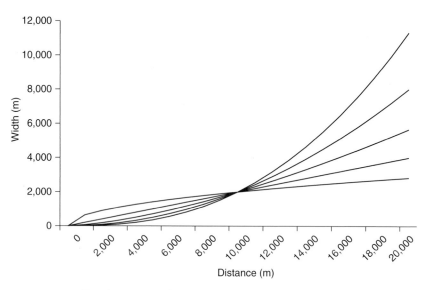

FIGURE 3.2   Estuarine widths after Eq. 3.1, for $\lambda = 10,000$, $W_L = 2,000$ and for $n = 0.5, 1.0, 1.5, 2.0$ and 2.5 respectively (lower to upper traces right hand margin) (Toolbox 2).

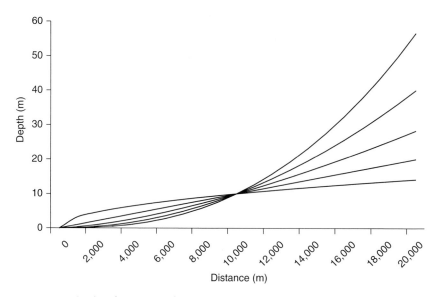

FIGURE 3.3   Estuarine depths after Eq.3.2, for $\lambda = 10,000$, $D_L = 10$ and for $m = 0.5, 1.0, 1.5, 2.0$ and 2.5 respectively (lower to upper traces right hand margin) (Toolbox 3).

from mouth (m), a is a coefficient, and the subscript zero denotes values at the mouth of the estuary.

In a similar analysis, Wright et al. (1973) have also shown that depth can be represented by

$$D_x = D_0 e^{-b(x/L)} \qquad \text{Eq. 3.6}$$

where D_x is the estuary depth (m), L is the estuary length (m), x is the distance from mouth (m), *a* and *b* are coefficients, and the subscript zero denotes values at the mouth of the estuary.

This approach is carried forward into the following chapter as a basis for the quantitative description of generalized estuarine bathymetry.

3.6 EQUILIBRIUM CROSS-SECTION

There have been a number of attempts to determine the equilibrium form of estuarine bathymetry. For example, the hypsographic curve, which is a plot of the percentage of elevation and depth distribution, has found some use in assessing the maturity of estuarine systems (e.g. Willgoose and Hancock, 1998; Boumans et al., 2002).

A related, purely geometrical result concerns the frequently remarked relationship between estuarine volumes and the cross-section of the estuary at different stations. For example, Coats et al. (1995) followed Chantler (1974) and studied salt marsh inlets in California and demonstrated a clear relationship between the tidal prism and the cross-sectional area of the entrance as shown in Figure 3.4. Dronkers (2005) expresses O'Brien's (1969) relationship as

$$A_c = C_A P^q \qquad \text{Eq. 3.7}$$

where A_c is the inlet cross-sectional area at the mouth, P is the tidal prism, and C_A and q are empirical coefficients. Jarret (1976)

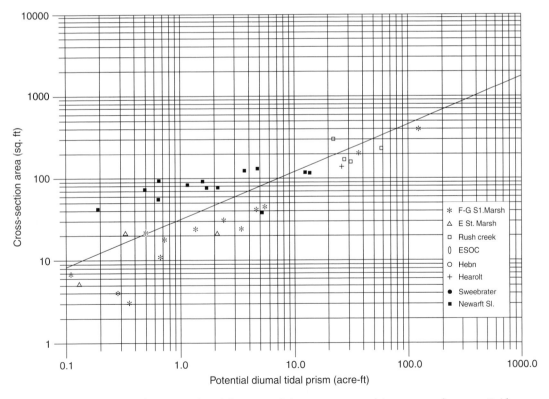

FIGURE 3.4 Relationship between inlet tidal prism and the cross-section of the entrance for some Californian examples (after Coats et al. (1995)).

fitted Eq. 3.7 to 162 inlets along the US coastline and derived values for q between 0.84 and 1.1. The coefficients may depend upon such factors as littoral drift, flood-ebb asymmetry, river discharge and the influence of headlands (Jarret, 1976; Hume and Herdendorf, 1992; Gao and Collins, 1994). Dronkers (2005) concludes that many inlets are fairly well represented by values of $C_A = \omega/2$ and $q = 1$ where ω is the radial frequency of the tide (rad s^{-1}).

3.7 ESTUARINE PLAN FORM

One measure of the plan form of an estuarine system is to consider the sinuosity of the channels which often develop in the upper reaches of macrotidal estuaries. For example, Hood (2004) compared historic and modern aerial photographs of three sites on Washington's Skagit River finding that dyke construction caused a loss of tidal channel surface area and a decrease in channel sinuosity.

3.8 BATHYMETRIC CHANGE

Bathymetric change in the Humber has been accretionary throughout the Holocene (Rees et al., 2000) at a rate of about 1,500,000 m^3 a^{-1}. Hardisty et al. (1998) constructed Digital Terrain Models (DTM) of the Humber Estuary based upon the techniques described by Petrie and Kennie (1987), Moore et al. (1991), and Lane et al. (1994). The DTMs were derived from Associated British Ports' charts of the estuary for 1980 and 1989 and were digitized at irregular spacings with additional data from the Admiralty Chart east of Spurn Head. The estuarine margin was digitized from local 1 : 500,000 Ordnance Survey maps and

used to generate pseudo depths set to 1 m above high water spring tides. In addition, the shoreline was used to constrain the DTM construction. The data were processed in the ARC-INFO Geographical Information System using the Delaumy triangulation method and sampled on a regular 100 m grid.

A number of important results were derived from the resulting DTMs. First, the water volumes in the estuary were estimated from the 1989 survey as:

	MHWS	**MHWN**	**MLWN**	**MLWS**
Ordnance Datum	3.9	1.8	−1.2	−3.0
Volume (km^3)	2.7	2.1	1.4	1.0

The increase in water volume within the estuary (effectively the tidal prism) was thus estimated at 1.7 and 0.7 km^3 for a spring and neap tide respectively. The estuary volume change was derived by comparison of the 1980 and 1989 DTMs for four ranges of levels:

From (m, OD)	To (m, OD)	Change km^3
−20	−10	Accretion of 0.02
−10	−5	No change
−5	+2	Erosion of 0.015
above +2		No change

The data show that, overall, there was accretion (principally in the deeper channels) of 0.005 km^3 between 1980 and 1989. The figures represent annual accretion of 0.56×10^6 m^3 or approximately 10^6 tonnes (1 MT) of sediment which is similar to the figure derived from the sedimentological work by Rees et al. (2000) and discussed above. This is equivalent to about 800 m^3 of accretion on an average tide and, therefore, a tidal influx of some 1280 T of sediment. Finally, since the flood and ebb transports on a typical tide are some 100–150,000 T, then it is evident that the very small residual figure is particularly difficult to determine or to detect in practice as discussed by Dyer et al. (2001).

3.9   SUMMARY

The flaring, trumpet-shaped width of many estuaries can be represented by:

$$W_x = W_0 e^{-a(x/L)} \qquad \text{Eq. 3.8}$$

where W_x is the estuary width (m), L is the estuary length (m), x is the distance from mouth (m), a is a coefficient, and the subscript zero denotes values at the mouth of the estuary.

The depth can be represented by:

$$D_x = D_0 e^{-b(x/L)} \qquad \text{Eq. 3.9}$$

where D_x is the estuary depth (m), L is the estuary length (m), x is the distance from mouth (m), a and b are coefficients, and the subscript zero denotes values at the mouth of the estuary.

Many estuaries have been accreting throughout the Holocene and, in the Humber, the accretion rate is approximately $1 \, \text{MT} \, \text{a}^{-1}$.

4

MODELING BATHYMETRY

Contents

4.1 INTRODUCTION

The estuarine processes described here are examined in some detail through the construction of a numerical model in each of the subsequent sections of the book. The model is intended as a research tool and additional experiments are described in the appropriate chapters. The model has also proven useful in graduate programs and its construction and testing is described in detail. The code at each stage is included on this book's web site. Following Hardisty et al. (1993), the modeling utilizes Microsoft Excel running in Windows.

The more specialized field of modeling particular aspects of estuarine dynamics is covered in a number of research papers, e.g. Bai et al., 2003.

4.2 BACKGROUND INFORMATION

In this section the steps involved in opening Microsoft Excel are described, text is entered, and cells are formatted. There are five stages in the work:

1 Locate the Excel icon and double click with the left mouse button.
2 Excel will open and may include many toolbars and buttons. The idea presently is to keep things simple in order to focus on estuary modeling. Open the View menu and click off all Toolbars except for the Formula Bar.
3 Point with the cursor and double click the Sheet 1 tab and rename it "main model."
4 Point to and press the "whole sheet" button located to the left of the column header A

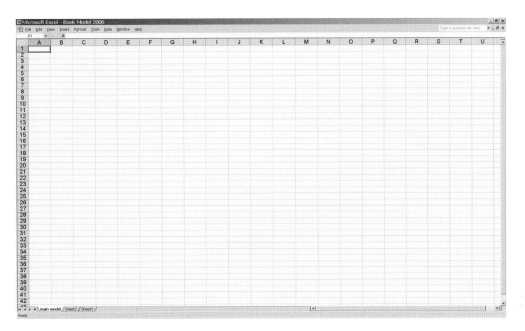

FIGURE 4.1   The formatted screen in Microsoft Excel.

and above the row 1. Choose the Format Cells Font dialog box and select Arial 8 point for the full spreadsheet and press ok. Use the Zoom command to set up the view so that Row 42 is at the bottom. Select Format Column Width and set all columns so that Column U is to the right. Click ok.

5 Save the workbook with an appropriate name as in Figure 4.1.

4.3   SETTING OUT THE ESTUARY MODEL

In this section we set out the estuary model's main screen. There are three stages in this part of the work:

1 Select K2 and enter a title such as "Humber Estuary Numerical Model" clicking the tick to enter the new text. Set the new text to 14 point, bold, dark blue, and centralized in the Format Cells dialog box.

2 Select and copy the map from the book's web site and position it over G3:O21.

3 Finally enter "CONTROL PANEL" into B3 on the main model sheet in bold dark blue and click ok to complete this section of the work as shown in Figure 4.2.

4.4   DEFINING THE ESTUARY

Although the model estuary is necessarily a fabrication or simulation, it is useful to base the work upon a real estuary. In this book, many results are reported from the Humber which is the largest estuary in England and drains about 25,000 km^2 into the North Sea as detailed in Chapter 1. For the present purposes we therefore choose some of the estuary's towns and villages to give life to the synthetic model. There are five short stages in this section of the work:

1 Select Sheet 2 and rename the sheet tab to bathymetry. Select and set the whole sheet to Arial 8 point. Adjust the column width and zoom until Row 42 and Column U make

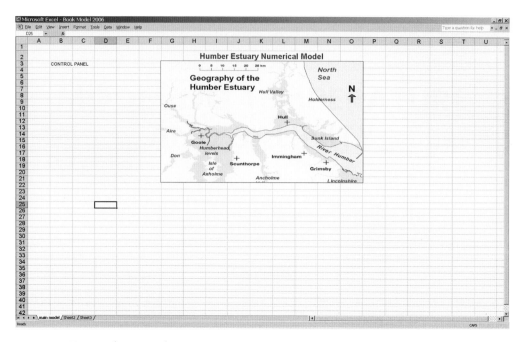

FIGURE 4.2   Entering the text and image.

up the bottom and right-hand side of the display.

2 Select   C28 and enter   Goole
           E28   Trent Falls
           G28   Brough
           I28   Bridge
           K28   Kingston upon
                             Hull
           O28   Immingham
           Q28   Grimsby
           S28   Spurn Head

3 Multiple select C28:S28 and use the Format Cells Align dialog box to set horizontal center and click ok.

4 Select   A29 and enter   Kms from Mouth
           A30   Width m
           A31   Depth m
           A32   Cross section m^2
           A33   Vol upstream
                             m^3 × 10^6

Superscripts are obtained by selecting the characters in the formula bar and font formatting with the superscript pip.

5 Finally enter 80 into C29, 75 into D29, and so on until you enter 0 into S29. Center align C29:S29. The resulting screen is shown in Figure 4.3.

4.5   MODELING ESTUARINE WIDTH

In the present section the estuarine width is modeled according to the formula developed in Chapter 3:

$$W_x = W_0 e^{(-ax/L)} \qquad \text{Eq. 4.1}$$

where W_x is the width at x, W_0 is the width at the estuary mouth, e is the exponential constant, 2.7, a is the width coefficient, x is the distance from the mouth, and L is the length of the estuary.

There are five stages in this section of the work:

1 Select C30 and type in the formula

  =12000*2.7^(−4.25*C29*1000/80000)

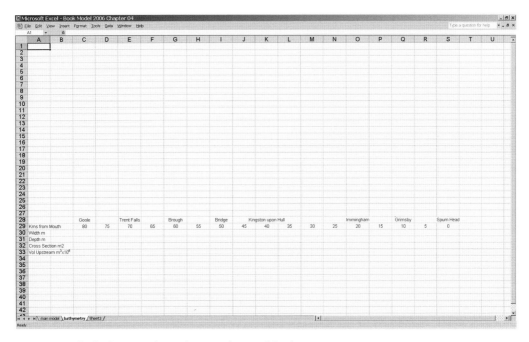

FIGURE 4.3 The bathymetry sheet after completion of the five stages.

which can be read as follows:

- the equality sign signifies that a formula follows
- W_0 is the width of the estuary mouth and is 12,000 m in the present example
- the asterisk sign represents multiplication in Excel
- e is the exponential or base of natural logarithms which is 2.7
- the tilde sign (upper case 6 or ˆ) represents raise to the power in Excel
- the coefficient a in Eq. 4.1 is 4.25 which is negative
- C29 contains the distance from the mouth of the estuary, x in Eq. 4.1
- which must be multiplied by 1000 to change from kilometers to meters
- / represents division in Excel and, in this case, it is division by the length of the estuary, L in Eq. 4.1, which in our example is 80 km represented by 80,000 m.

2 Click the tick in the formula bar to enter the formula into the cell. Excel calculates and displays the result which at the head of the estuary some 80 km inland is 176.1511.

3 Presently, however, the cell has the default free-floating display format and we require integers and to use a comma to indicate the thousands. With C30 selected, choose Format Cells dialog box, center the display in Alignment and, in Number, choose zero decimal places and the comma thousand. Click ok as shown in Figure 4.4.

4 Finally, select C30 to S30. Choose Edit Fill Right and the formula complete with relative address correction is duplicated into all selected cells.

The estuary width model is now complete and shows an exponential decrease from 12,000 m at Spurn through 1,454 m at Kingston upon Hull and down to 176 m at Goole. These widths are very similar to the real Humber Estuary that was described in Chapter 1. The result is shown in Figure 4.5.

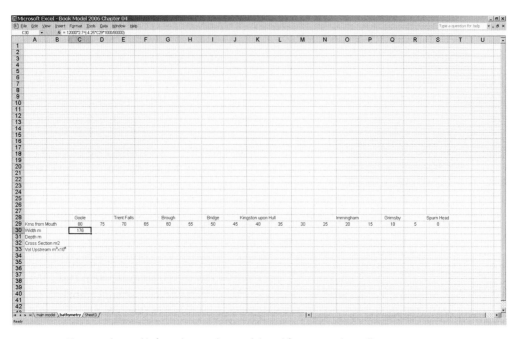

FIGURE 4.4   Entering the width formula into the model and formatting the cell.

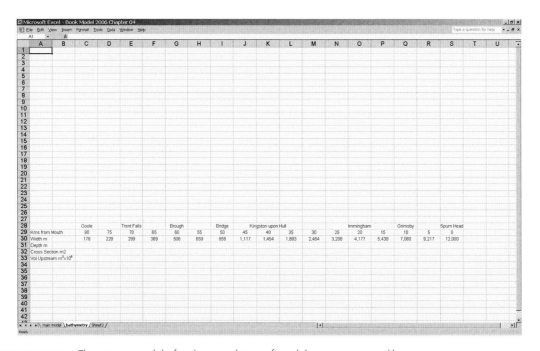

FIGURE 4.5   The estuary model after the completion of modeling estuarine width.

4.6 MODELING ESTUARINE DEPTH AND CROSS-SECTION

The estuarine depth decreases exponentially according to the following formula (Chapter 3):

$$D_x = D_0 e^{(-bx/L)} \qquad \text{Eq. 4.2}$$

where, to recap, D_x is the depth at x, D_0 is the depth at the estuary mouth, e is the exponential constant, 2.7, b is the depth coefficient, x is the distance from the mouth, and L is the length of the estuary.

The estuarine depths are modeled using this formula in the present. There are five stages in the work:

1 Select C31 and type in the formula:

=15*2.7ˆ(−1.4*C29*1000/80000) Eq. 4.3

which can be read as follows:

- the equality sign tells Excel that a formula follows
- D_0 is the depth of the estuary mouth and is 15 m in the present example
- the asterisk sign represents multiplication in Excel
- e is the exponential or base of natural logarithms which is 2.7
- the tilde sign (upper case 6 or ^) represents raise to the power in Excel
- the coefficient b in Eq. 4.2 is 1.4 which is negative
- C29 contains the distance from the mouth of the estuary, x in Eq. 4.2
- which must be multiplied by 1000 to change from kilometers to meters
- / represents division in Excel and, in this case, it is division by the length of the estuary, L in Eq. 4.2. In the Humber this is 80 km represented by 80,000 m.

2 Make sure that you understand the formula and then press the tick in the formula bar to enter the formula into the cell. Excel immediately performs the calculation and displays the result which at the head of the estuary is 3.734.

3 Presently, however, the cell has the default free-floating display. With C31 selected, choose Format Cells, center the display and use the Number dialog box to choose one decimal place.

4 Select C31 to S31. Choose Edit Fill Right and the formula complete with relative address correction is duplicated into all selected cells.

5 The estuary depth model is now complete and should show an exponential decrease from 15.0 m at Spurn through 7.5 m at Kingston upon Hull and down to 3.7 m at the head of the estuary at Goole. These depths are very similar to the real Humber Estuary which was described in Chapter 1. The result is shown in Figure 4.6.

The estuary's cross-sectional area is easily calculated from the product of the width and depth at each station according to the following formula:

$$A_x = W_x D_x \qquad \text{Eq. 4.4}$$

where A_x is the cross-sectional area in m^2 and W_x and D_x are the width and depth respectively at a distance x from the mouth. The cross-section model is entered into the spreadsheet in three stages:

1 Select C32 and enter =C30*C31 which can be read as the product of the width contained in C30 and the depth contained in C31.

2 Use the number dialog box in the Format Cells menu to select zero decimal places and the thousand comma with center alignment and click ok.

3 Finally select C32 to S32 and Fill Right to duplicate the formula along the estuary.

The estuarine cross-section model is now complete and shows an exponential decrease from 180,000 m^2 at Spurn through 10,881 m^2 at Kingston upon Hull and down to 658 m^2 at Goole as shown in Figure 4.7.

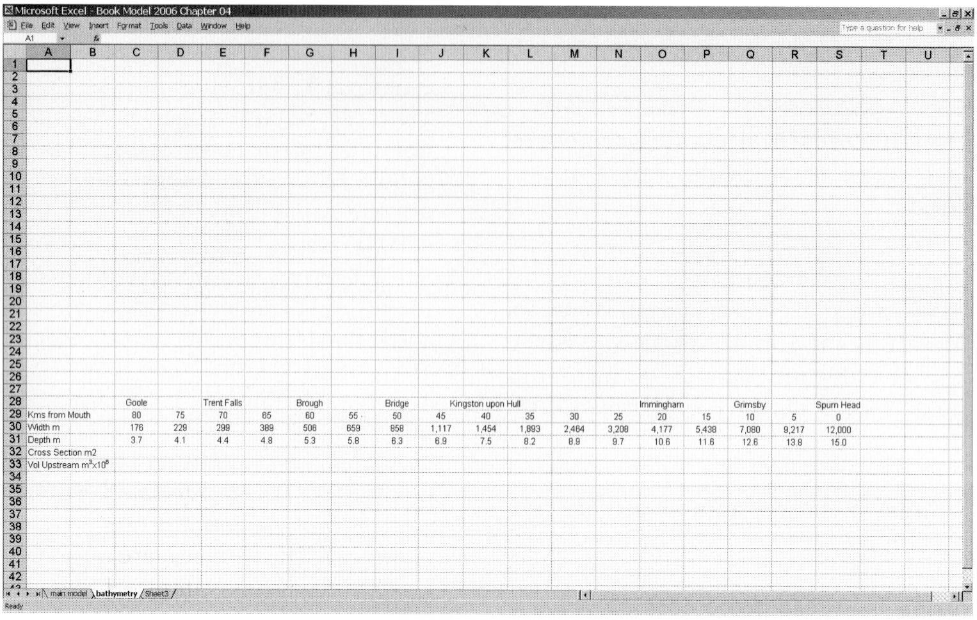

		Goole		Trent Falls		Brough		Bridge		Kingston upon Hull						Immingham		Grimsby		Spurn Head	
Kms from Mouth		80	75	70	65	60	55	50	45	40	35	30	25	20	15	10	5	0		
Width m		176	229	299	389	506	659	858	1,117	1,454	1,893	2,464	3,208	4,177	5,438	7,080	9,217	12,000		
Depth m		3.7	4.1	4.4	4.8	5.3	5.8	6.3	6.9	7.5	8.2	8.9	9.7	10.6	11.6	12.6	13.8	15.0		
Cross Section m2																				
Vol Upstream m³x10⁶																				

FIGURE 4.6 The estuary model after the completion of the width and depth modeling.

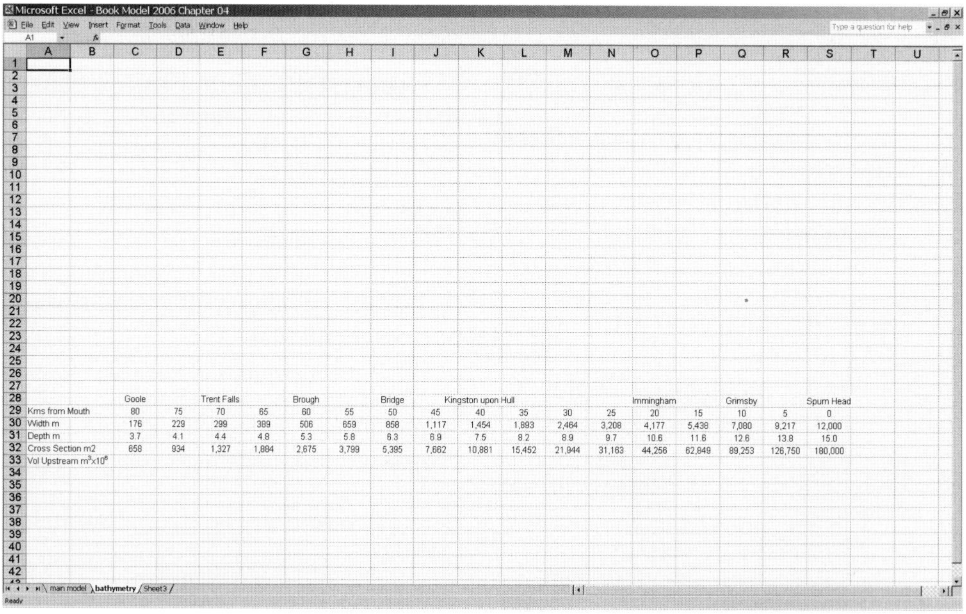

		Goole		Trent Falls		Brough		Bridge		Kingston upon Hull						Immingham		Grimsby		Spurn Head	
Kms from Mouth		80	75	70	65	60	55	50	45	40	35	30	25	20	15	10	5	0		
Width m		176	229	299	389	506	659	858	1,117	1,454	1,893	2,464	3,208	4,177	5,438	7,080	9,217	12,000		
Depth m		3.7	4.1	4.4	4.8	5.3	5.8	6.3	6.9	7.5	8.2	8.9	9.7	10.6	11.6	12.6	13.8	15.0		
Cross Section m2		658	934	1,327	1,884	2,675	3,799	5,395	7,662	10,881	15,452	21,944	31,163	44,256	62,849	89,253	126,750	180,000		
Vol Upstream m³x10⁶																				

FIGURE 4.7 Screen display after the completion of the above mentioned three stages.

4.7 GRAPHICAL DISPLAY

Graphs of the estuary's width, depth, and cross-sectional variations are plotted in this section. There are seven stages in this work:

1 Multiple select C28:S28 and press the CTRL key. Release the mouse button, keep the CTRL key depressed, and multiple select C30:S32 as shown in Figure 4.8.
2 Select Insert Chart from the menus and then follow the Step Wizard. On Step 1 of 4 select the top left line chart and click next.
3 On Step 2 of 4 check that the data are in rows and click next.
4 On Step 3 of 4 type in a title, remove the legend, and click next.
5 On Step 4 of 4 select as an object in bathymetry and click Finish.
6 Locate the chart on the spreadsheet between B2 and T26. Only two plots are immediately visible, because the depth data are too small. Double click on the *y*-axis and select logarithm scale.
7 For a black and white printer, double click on each graph line in turn and change the color to black. The resulting plot is shown in Figure 4.9.

4.8 MODEL VALIDATION

The width and charted depth of the estuary were determined from the British Transport Docks Board Chart (1977) and Gameson (1982). These are compared in Table 4.1 with the values obtained in the model.

Some inconsistencies are apparent. For example, the low water width is from the tip of Spurn Head to the south bank whereas the high water width includes Spurn Bight. Further upstream, the chalk Wolds cause narrowing at the Bridge. Also, charted depths are approximations to deepest values

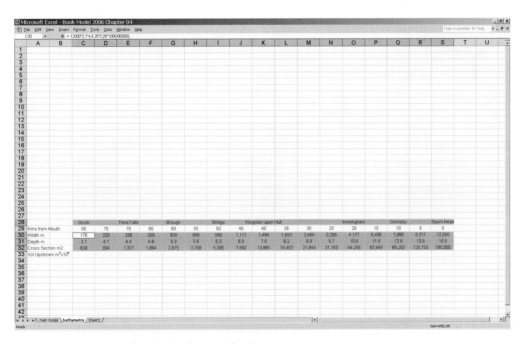

FIGURE 4.8 Disconnected multiple selection of cells.

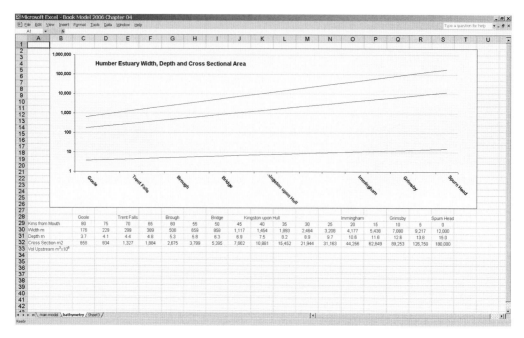

FIGURE 4.9 Displaying the estuary's width, depth, and cross-section.

TABLE 4.1 Comparison of modeled and charted widths and depths for the Humber Estuary.

Distance	Location	Chart low water width (km)	Chart high water width (km)	Charted depth[3]	Model width (km)	Model depth (m)
0	Spurn[1]	7.0	12.2	14.0	12.0	15.0
10	Grimsby	5.0	5.8	11.0	7.1	12.6
20	Imm'ham	3.2	4.2	10.0	4.2	10.6
40	K upon Hull	1.8	2.6	8.0	1.5	7.5
50	Bridge[2]	1.7	2.4	4.1	0.9	6.3
60	Brough	2.0	2.6	3.0	0.5	5.3
70	Trent Falls	0.3	0.6	2.0	0.3	4.4
80	Goole	0.1	0.2	2.0	0.2	3.7

[1] From the tip of Spurn Point to Tetney.
[2] Between the bridge piers.
[3] Mean charted depth.

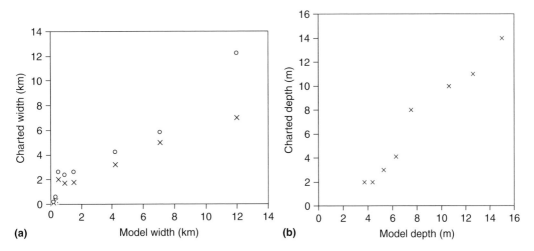

FIGURE 4.10   (a) Model widths (km) versus low water widths (X, km) and high water widths (km) and (b) Model depths (km) versus charted depths (m) for the Humber (Toolbox 4.8.1).

shown in main channels with respect to Chart Datum (Chapter 3).

The modeled and observed results are compared in Figure 4.10a for the widths and Figure 4.10b for the depths. There is clearly a reasonable agreement between the modeled results and the values obtained from the charts, particularly when the high water widths and depths are utilized.

Part III

TIDES IN ESTUARIES

5

ESTUARINE TIDES

Contents

5.1 INTRODUCTION

This chapter deals with the generation of tides in the deep water of ocean basins and with tidal propagation into the shallow water of estuaries. The chapter is based on Huntley (1979), Pugh (1987, 2004), and Hardisty (1990a). We define a tide as

> A periodic vertical or horizontal water movement, which has a coherent amplitude and phase relationship to some periodic geophysical movement.

The dominant geophysical movement is due to the variation in the gravitational field at the surface of the Earth caused by the Earth–Moon and Earth–Sun systems. Movements due to the Moon and the Sun are termed "gravitational tides" to distinguish them from movements induced by periodic atmospheric forcing, termed "meteorological tides." The result is that estuarine water levels rise and fall in a regular and predictable manner. In this chapter, we derive the equations for the frequency and magnitude of such tidal water level changes.

5.2 BACKGROUND INFORMATION

There are a number of coastal texts which introduce the material in this chapter. For example, Woodroffe (2003) or Dronkers (2005) have qualitative descriptions of tidal theory and there are a number of excellent Internet resources available online. The chapter is based on quantitative descriptions of tidal phenomena including the now classical Doodson and Warburg (1941), Neumann and Pierson (1966), and Pugh (1987, 2004), along with other more specific references cited in the body of the text. Basic tidal theory is detailed and, in particular, explanations and formulae are sought for the manner in which the tide develops and changes as it enters the estuarine environment. The techniques for monitoring the tides were covered in Chapter 2 and the equations which are developed here are incorporated into the numerical model in the following chapter.

5.3 A BRIEF HISTORY OF TIDAL THEORY

There is little doubt that coastal dwellers have always been aware of tidal phenomena. Excavations in the Indian district of Ahmedabad, for example, have revealed a tidal dockyard which dates back to 2,450 BC. Aristotle, in the fourth century BC, wrote that the "ebbings and risings of the sea always come around with the Moon and upon certain fixed times." At about the same time, Pytheas traveled through the Straits of Gibraltar to the British Isles, where he observed large tides at twice-daily periods. He is said to have been the first to report the half-monthly variations in the range of the Atlantic Ocean tides, and to note that the greatest ranges, which we call spring tides, occurred during the new and full Moon. He also recorded the strong tidal streams of the Pentland Firth between Scotland and Orkney.

The ideas advanced by early philosophers to explain the connection between the Moon and the tides were, at first, speculative. Arabic writers supposed the Moon's rays to be reflected off rocks at the bottom of sea, thus heating and expanding the water, which then rolled in waves toward the shore. One poetic explanation invoked an angel who was set over the seas: when he placed his foot in the sea the flood of the tide began, but when he raised it, the tide ebbed. In about AD 730, the Northumbrian monk Bede, described how the rise of the water along one coast of the British Isles coincided with a fall elsewhere. Bede also knew of the progression in the time of high tide from north to south along the English east coast. By the mid-seventeenth century, understanding was progressing beyond the descriptive approaches and toward the resolution of three different theories:

1 Galileo (1564–1642) proposed that the rotations of the Earth, annually around the Sun and daily about its own axis, induced motions of the sea that were modified by the shape of the seabed to give the tides.
2 The French philosopher Descartes (1596–1650) thought that space was full of invisible matter known as ether, and the movement of the Moon round the Earth compressed the ether in a way which transmitted pressure to the sea, hence forming the tides.
3 Kepler (1571–1630) was one of the originators of the idea that the Moon exerted a gravitational attraction on the waters of the ocean, drawing it toward the place where it was overhead. This attraction was balanced by the Earth's attraction on the waters for "If the earth should cease to attract its waters, all marine waters would be elevated and would flow into the body of the Moon."

Gradually, as the concept of a heliocentric system of planets became established, and the laws of the planetary motion and gravitational attraction were developed, Kepler's ideas became increasingly accepted and were finally formalized by Isaac Newton (1642–1727). Newton made a major advance in the scientific understanding of the generation of tides, as well as many other phenomena. He used tidal motions in his formulation of the law of gravitational attraction:

> two bodies attract each other with a force which is proportional to the product of their masses and inversely proportional to the square of the distance between them.

Newton's gravitational theory also explained the twice-monthly spring to neap cycle, diurnal inequalities, and equinoctial tides. Newton's work became known as the equilibrium theory of tides and remains the basis for our modern understanding of tidal phenomena.

5.4 EQUILIBRIUM THEORY OF TIDES

The analysis of tidal motions divides into two approaches and both are detailed in this chapter. Newton's equilibrium theory of tides is described in this section and provides the theoretical explanation for the range and period of ocean tides. Once generated, however, these tides propagate into shallow water and are deflected and reflected causing the ocean tide to change before reaching the estuary. The equilibrium theory does not, however, provide the site-specific predictions which are required in estuarine models. Instead, the alternative approach, known as harmonic analysis, is more useful and is described in the following section.

Newton showed that the force of attraction between two bodies M_1 and M_2 was proportional to the product of their masses and inversely proportional to the square of their separation r. Newton defined the constant of proportionality as G, the universal gravitational constant ($6.672 \times 10^{-11} \, \mathrm{N \, m^2 \, kg^{-2}}$), so that the force is:

$$F = G\frac{M_1 M_2}{r^2} \qquad \text{Eq. 5.1}$$

Consider a simple, spherical Earth with a layer of water (Figure 5.1). Each of the water masses experiences a tide generating force, F_T, equal to the difference between the gravitational attraction exerted on that mass by the Moon, F_G, and the centrifugal force, F_C, due to the orbit of the Earth and Moon around a point of rotation which is actually inside the Earth at a distance of 4,600 km from its center.

$$F_T = F_G - F_C \qquad \text{Eq. 5.2}$$

Consider a particle of mass m located at P_1 in Figure 5.1. The centers of the Earth and the Moon are separated by a distance R_m, and the Earth's radius is a. The particle experiences a gravitational pull toward the Moon from Eq. 5.1 of

$$F_G = \frac{GmM_m}{(R_m - a)^2} \qquad \text{Eq. 5.3}$$

where M_m is the mass of the Moon. The particle experiences a centrifugal force

$$F_C = \frac{GmM_m}{R_m^2} \qquad \text{Eq. 5.4}$$

The difference between these two forces is the tide generating force which is directed

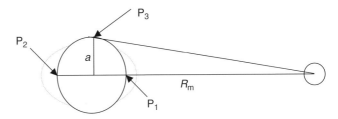

FIGURE 5.1 Definition sketch for the determination of the equilibrium tide.

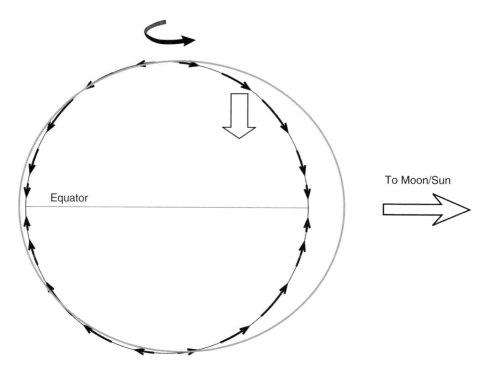

FIGURE 5.2 Global distribution of the tide generating force.

toward the Moon at P_1 and hence has zero tangential components (Figure 5.2). Thus the tidal force has a value of approximately

$$F_T = \frac{Gm2aM_m}{R_m^3}$$ Eq. 5.5

This force, at equilibrium, is balanced by a positive water slope so that the ocean bulges toward the Moon at P_1. Similar expressions can be obtained for the distribution of the tide generating force at other locations, such as at P_2 where the force is directed away from the Moon generating a second bulge as shown in Figure 5.2. The result is that the water is drawn up into two bulges at equilibrium: one beneath the Moon and a second on the opposite side of the planet. As the observer (e.g., at the arrowed location in Figure 5.2) rotates, whilst the planet spins the two bulges are experienced as high waters separated by

two low waters. Since it takes 24.42 hours for the Earth to rotate beneath the Moon (which orbits slightly and is thus further advanced after each rotation), these two tides are experienced during each orbit, known as the lunar, semi-diurnal, or M_2 tide (Figure 5.3a).

Similar relationships may be established for the attractive forces due to the Sun, which are equivalent to the numerical values obtained from Eq. 5.5 multiplied by a factor of 0.46 to take account of the different values of mass and orbital diameter. Two bulges due to the Sun are experienced in every 24-hour period, known as the solar semi-diurnal tide, S_2. If the Moon is not orbiting above the Earth's equator then its two bulges will be centered at different latitudes resulting in the diurnal inequality frequently observed in tides (Figure 5.3b).

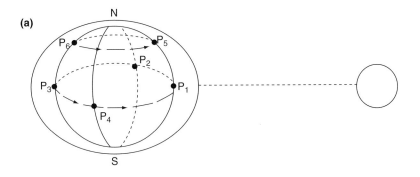

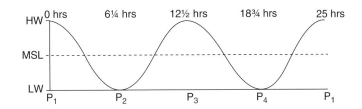

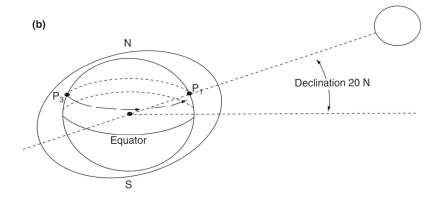

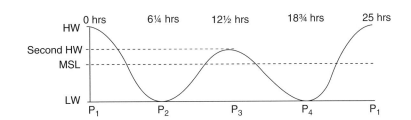

FIGURE 5.3 The generation of a single tide, here the M_2 or S_2 tide as a sinusoid resulting from the rotation of the observer beneath the two gravitational bulges; (a) represents the symmetrical tides generated by the Moon orbiting above the Earth's equator whilst (b) represents the diurnal inequality due to an inclination of the orbit (from DIR 5.1).

5.5 HARMONIC ANALYSIS OF TIDES

It is seldom possible to derive the full theoretical expression for tidal heights in estuaries because of the shallow water and coastal effects. It is more useful to utilize harmonic analysis to derive the main constituents and then to express the resultant as a summation of sinusoidal contributions. Harmonic analysis involves the decomposition of the tide into a series of regular sinusoidal constituents of a given or determinable period and amplitude and then the formulation of the result by simple addition. The constituents represent the influence of the gravitational forces postulated by Newton's theory but their individual parameterization is performed from an analysis of the local tidal measurements and the result is site specific. In general, the height of the tide at a site, $h(t)$, is given by this addition:

$$h(t) = A_1 \cos(\omega_1 t - \rho_1) + A_2 \cos(\omega_2 t - \rho_2) + \cdots$$

Eq. 5.6

where A_n represents the amplitude, ω_n the angular velocity (usually expressed in degrees per solar hour), t is the time, and ρ_n accounts for phase differences between the terms. Various almanacs and the British Admiralty's Manual of Tides provide details of the many constituents currently used for tidal analysis, with mean values for speed numbers and coefficients for the equilibrium tide.

A Scotsman, Lord Kelvin, devised the method of harmonic prediction around 1867. He also invented the first of a series of machines for tidal height predictions with rotating pulleys set on shafts (Figure 5.4). The size and angle of the pulley on each shaft is determined from the harmonic constants. There are a number of shafts, each one corresponding to a different frequency component of the tide raising effect. A system of pulley blocks and wires are arranged to add up the effect from the many shafts and to produce the tidal curve. The first machine

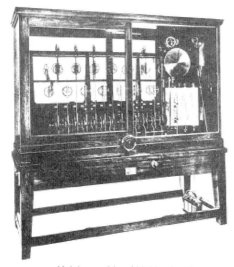

Kelvin machine (tidal institute)

FIGURE 5.4 The original Kelvin and the later Doodson/Lege tide prediction machines (DIR 5.2).

provided for the summation of 10 of the principal constituents. Without such machines the harmonic method of prediction was simply too labor intensive to be useful.

5.6 HARMONIC TERMS

The most important frequencies and potential amplitudes of the major components are given in Table 5.1, although not all of the components have to be considered at a particular location. Indeed, the estuary model is developed using only three components in the following chapter.

Each of the constituents propagates through coastal waters and into the estuarine environments. For example, Figure 5.5 shows the tidal chart for the lunar, semi-diurnal tide in the North Sea. The tide rotates in an anticlockwise direction around three amphidromic points, one to the southwest of Norway, a second in the central German Bight, and a third off Suffolk in the southern North Sea. The location of the tide at hourly intervals is shown by the co-tidal lines which radiate out from the amphidromes. The range of the tide increases from near-zero at the amphidromic points, as shown by the contours or co-range lines which encircle the amphidromes.

Pugh (2004) represents the lunar and solar, semi-diurnal tides as

$$h_{S2}(t) = A_{S2} \cos(2\pi t/T_{S2} + \rho_{S2}) \qquad \text{Eq. 5.7}$$

$$h_{M2}(t) = A_{M2} \cos(2\pi t/T_{M2} + \rho_{M2}) \qquad \text{Eq. 5.8}$$

where $h_{S2}(t)$ and $h_{M2}(t)$ are the depth of water at time t, A_{S2} and A_{M2} are the amplitudes, T_{S2} and T_{M2} are the periods, and, ρ_{S2} and ρ_{M2} are the phases, due to the

TABLE 5.1 *Major tidal potential constituents.*

Symbol	Name	Period (h)	Speed (h)	Coefficient
Semi-diurnal components				
M_2	Principal lunar semi diurnal	12.42	28.9841	0.908
S_2	Principal solar semi diurnal	12.00	30.0000	0.423
N_2	Larger lunar elliptic	12.66	28.4397	0.174
K_2	Luni-solar semi diurnal	11.97	30.0821	0.115
ν_2	Larger lunar evectional	12.63	28.5126	0.033
μ_2	Variational	12.87	27.9682	0.028
L_2	Smaller lunar elliptic	12.19	29.5285	0.026
T_2	Larger solar elliptic	12.01	29.9589	0.025
$2N_2$	Lunar elliptic second order	12.91	27.8954	0.023
Diurnal components				
K_1	Luni-solar diurnal	23.93	15.0411	0.531
O_1	Principal lunar diurnal	25.82	13.9430	0.377
P_1	Principal solar diurnal	24.07	14.9589	0.176
Q_1	Larger lunar elliptic	26.87	13.3987	0.072
M_1	Small lunar elliptic	24.86	14.4921	0.040
J_1	Small lunar elliptic	23.10	15.5854	0.030
Long period components				
M_f	Lunar fortnightly	327.8	1.0980	0.156
M_m	Lunar monthly	661.3	0.5444	0.083

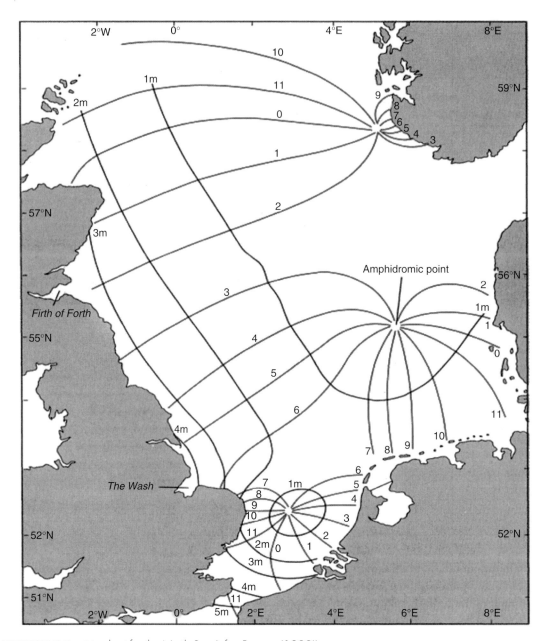

FIGURE 5.5 M_2 chart for the North Sea (after Brown, (1999)).

solar semi-diurnal and the lunar semi-diurnal tides respectively. The actual water depth, $h(t)$ at time t is then the numerical summation of the corresponding depths about a datum, z_0:

$$h(t) = h_{S2}(t) + h_{M2}(t) + z_0 \qquad \text{Eq. 5.9}$$

The lunar cycle at 29.5 days is approximately 1.035 times the solar day (Pugh, 2004) that is 24.84 hours and thus the period of the lunar

TABLE 5.2   Principle component for the Humber.

Station	Z_0	M_2	p_{M2}	S_2	p_{S2}	K_1	O_1	F_4	p_{F4}
Spurn Head	4.09	2.13	151	0.72	199	0.16	0.17	0.002	070
Immingham	4.18	2.29	163	0.76	213	0.15	0.18	0.004	196
K upon Hull	4.10	2.40	169	0.78	226	0.17	0.16	0.010	270
Bridge	3.59	2.33	174	0.80	228	0.14	0.16	0.026	292
Blacktoft	–	2.05	202	0.60	254	0.11	0.14	0.119	305

(a)

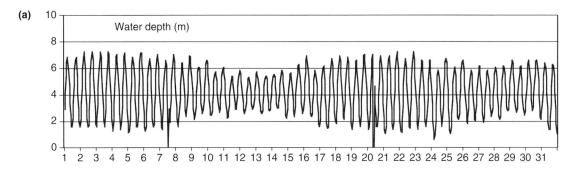

(b)

FIGURE 5.6   (a) Observed water depths at Immingham on the Humber Estuary for January 2003 and (b) phases of the Moon for the month and the actual view on January 10th (author's data).

semi-diurnal tide is 12.42 hours and that of the lunar quarter-diurnal tide is 6.21 hours.

Estuarine environments can be classified according to the relative importance of the amplitudes of four major terms as expressed by the Tide Form Number, F:

$$F = (K_1 + O_1)/(M_2 + S_2) \qquad \text{Eq. 5.10}$$

where O is the lunar diurnal and K is the soli-lunar diurnal. F is calculated for sites in the Humber from data in Admiralty Tide Tables (1996) in Table 5.2. Van Rijn (1990) notes that values of the tide form number

of less than 0.25 correspond to semi-diurnal tidal dominance which, clearly, applies to the Humber.

5.7   SPRING-NEAP VARIATIONS

The tides at Immingham in the Humber are shown for January 2003 (Figure 5.6a) and show that there are usually two high and two low tides each day and that the tide increases for the first few days to a maximum

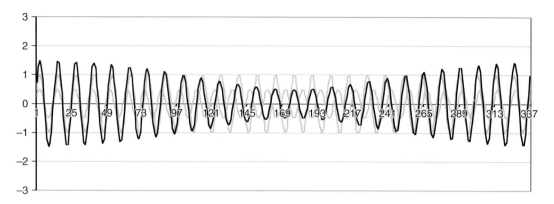

FIGURE 5.7   Spring-neap variation in the tidal range for $M_2 = 1.0$ and $S_2 = 0.5$ and periods of 12.42 and 12 h respectively (Toolbox 5.4.7).

on about the third of the month. It then decreases until about the 14th and builds up to an even greater range around about January 22nd. The largest tides are called the spring tides and occur near the times when the Moon and Sun are either in opposition (on opposite sides of the Earth) or in conjunction (on the same side of the Earth). The smaller tides are called neap tides and occur shortly after the first and third quarters of the Moon when the Moon is in quadrature (lines between the center of the Earth and the centers of the Moon and the Sun are roughly perpendicular).

The changes in the tidal range from neap to spring tides are explained by the beat effect of the M_2 and S_2 tides which reinforce at springs and are in opposition at neaps. Mathematically, the phase shifts (Eq. 5.7 and Eq. 5.8) are zero during springs and a maximum of $\pi/2$ at neaps. Figure 5.7 performs the addition in Eq. 5.9 and the result is a quite realistic simulation of the spring-neap cycle.

5.8   TIDES IN ESTUARIES

Upon entering an estuary, the water becomes shallow and the tidal wave is observed to become more asymmetric as it travels upstream (Figure 5.8). The explanation for this increasing asymmetry lies in the fact that friction causes the wave to travel at a speed governed by the water depth (Pugh, 2004):

$$c = \sqrt{gh} \qquad \text{Eq. 5.11}$$

where c is the speed (celerity) of the tidal wave (m s^{-1}), g is the gravitational acceleration (9.81 m s^{-2}), and h is the water depth (m).

The tidal wave length (λ_T) is the product of the celerity and the tidal period (T s),

$$\lambda_T = cT \qquad \text{Eq. 5.12}$$

thus the celerity and the wavelength decrease as the tide moves from the deep ocean into shallow water. The result is that, within the estuary, the crests are in deeper water than the troughs and the crests tend to catch up with the troughs producing the asymmetry. In the simple case of the M_2 tidal wave, Pugh (2004) showed that an M_4 component (the lunar quarter-diurnal species) is generated as the tide progresses into shallow water. The M_4 has twice the frequency of

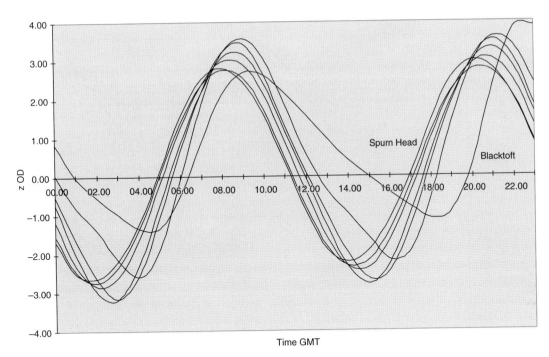

FIGURE 5.8 Increasing asymmetry upstream for the tidal wave in the Humber Estuary plotted from Spurn Head to Blacktoft.

the M_2 and amplitude given by

$$A_{M4} = \frac{3}{4}\frac{xA_{M2}^2}{\lambda_T h} = \frac{3}{4}\frac{xA_{M2}^2}{hT\sqrt{gh}} \qquad \text{Eq. 5.13}$$

Thus the amplitude of the M_4 increases as the distance increases along the channel. The amplitude of the quarter diurnal also increases if the channel depth is small, and as the square of the semi-diurnal component.

The resulting M_4 tide is therefore

$$h_{M4}(t) = A_{M4}\cos(2\pi t/T_{M4} + p_{M4}) \qquad \text{Eq. 5.14}$$

where $h_{M4}(t)$ is the depth of water at time t, A_{M4} is the amplitude (from Eq. 5.13), T_{M4} is the period, and p_{M4} is the phase.

The actual water depth due to the main harmonics, the lunar semi- and quarter-diurnal and the solar semi-diurnal constituents, $h(t)$ at time t is then the numerical summation of the corresponding depths about a datum, z_0. For spring tides this is

$$h(t) = h_{S2}(t) + h_{M2}(t) + h_{M4}(t) + z_0$$

$$= A_{S2}\cos\left(2\pi\frac{t}{12}\right) + A_{M2}\cos\left(2\pi\frac{t}{12.42}\right)$$

$$+ \frac{3}{4}\frac{xA_{M2}^2}{hT\sqrt{gh}}\cos\left(2\pi\frac{t}{6.21}\right) + z_0$$

Eq. 5.15

Figure 5.9a shows an implementation of Eq. 5.15 with $A_{M2} = 2.1$, $A_{S2} = 0.7$ (from Table 5.2) and $A_{M4} = 0$, and z_0 for the Humber of 4.1 m giving a tide with high water at about 7 m, low water at about 1 m and a tidal period of a little more than 12 h. The result compares reasonably well with the data in Figures 5.7 and 5.8.

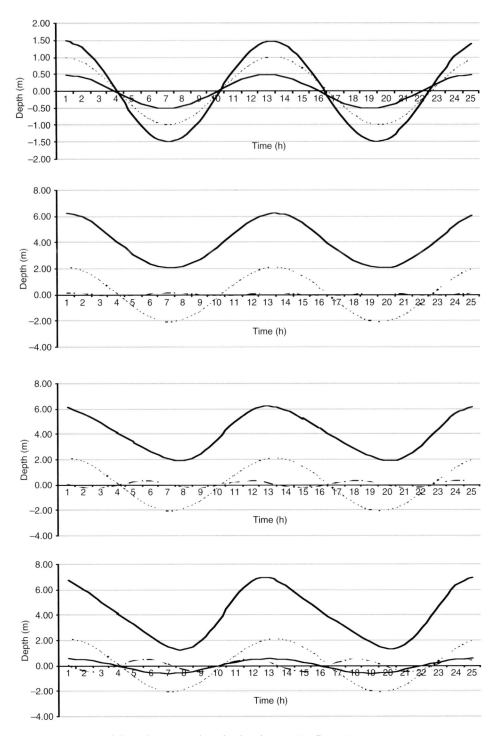

FIGURE 5.9   Estuarine tidal simulations as described in the text (Toolbox 5).

Figures 5.9b and 5.9c also implement Eq. 5.15 with the lunar tide, $A_{M2} = 2.0$ m, $A_{S2} = 0$, and $z_0 = 4.1$. Figure 5.9b shows the position at, for example, Blacktoft, $x = 80$ km (80,000 m) and depth $= 5$ m giving an A_{M4} of 0.34 m and with zero M_4 phase shift. Although this introduces some asymmetry, with a higher high water and a lower, low water, the tide remains symmetrical about high water which is not the effect of increased crest speed discussed above. Instead, an M_4 phase shift of 1.5 h ($\pi/2$ for the M_4) reproduces the required steep rise to high water and gentler fall to low water. Compare this with the Blacktoft results shown in Figure 5.8.

The three components are integrated by the introduction of the solar semi-diurnal with an amplitude of $A_{S2} = 0.6$ m at Blacktoft to conditions simulated in Figure 5.9c to produce the resulting addition of the three as shown. This compares reasonably well with the asymmetric, shallow water tide shown in Figure 5.8.

5.9 SUMMARY

This chapter has introduced the concepts of tidal theory including generation, the concept of an equilibrium tide, harmonic analysis, and the shallow water effects. The depth of water in an estuary can be described in terms of the S_2, M_2, and M_4 tides by Eq. 5.15:

$$h(t) = A_{S2} \cos\left(2\pi \frac{t}{12}\right) + A_{M2} \cos\left(2\pi \frac{t}{12.42}\right)$$

$$+ \frac{3}{4} \frac{xA_{M2}^2}{hT\sqrt{gh}} \cos\left(2\pi \frac{t}{6.21}\right) + z_0$$

Eq. 5.16

6

MODELING TIDES

Contents

6.1 INTRODUCTION

The preceding chapter has demonstrated that although tidal forcing is simply the result of the gravitational attraction of the Moon and the Sun on the waters of the world's oceans the resulting changes in water depths in estuarine environments are complex and far from simple. The tide rises and falls in a periodic and continuous fashion, the range of the tide varies along the estuary and on a daily basis through the spring-neap cycle, and the tide changes from being an almost symmetrical sinusoid at the estuary mouth to an asymmetrical curve in the headwaters. However, the harmonic approach suggests that some simplifications are possible. In this chapter, the estuarine tide is modeled on the assumption that the constantly changing water depths can be adequately represented by the addition of sinusoidal functions for the two main tidal species and by the growth of the lunar quarter-diurnal. The implementation of this model is described, and tested with comparisons to the Humber Estuary.

6.2 BACKGROUND INFORMATION

Chapter 5 demonstrated that the tide may be represented by the summation of the S_2 (solar semi-diurnal), the M_2 (lunar semi-diurnal), and the M_4 (lunar quarter-diurnal) tides. The actual water depth, $h(x, t)$ at a distance x from the mouth and time t is then the numerical summation of the corresponding depths about a datum, z_0:

$$h(x,t) = A_{S2}\cos\left(2\pi\frac{t}{1}\right) + A_{M2}\cos\left(2\pi\frac{t}{12.42}\right)$$

$$+\frac{3}{4}\frac{xA_{M2}^2}{h_x 12.42\sqrt{gh_x}}\cos\left(2\pi\frac{t}{6.21}\right) + z_0$$

$$\text{Eq. 6.1}$$

where A_{S2} and A_{M2} are the amplitudes of the solar and lunar semi-diurnal tide, h_x is the water depth at x, the constants 12, 12.42, and 6.21 represent the periods of the solar and lunar semi-diurnal and the lunar quarter-diurnal tides respectively, and g is the gravitational acceleration (≈ 9.81 m s^{-2}). It is noted that Eq. 6.1 assumes all three components in phase.

6.3   CONTROLLING TIDAL INPUTS

Text is input in this section to allow control of the tides within the estuarine model:

1 Open the model and immediately "save as" with a new version number.
2 Select the blank Sheet 3 and rename the sheet tab "spring-neaps." Select the whole sheet and set to centered Arial 8 point with a column width of 10. It may be necessary to set the zoom in order to achieve the same display size as below.
3 Return to the main model sheet and

Select B4	and enter	S_2 m
B6		M_2 m
B8		M_4 m
B10		Start Station
D8		Depth m
D9		Width m

4 The Toolbar is used to set up control icons. Choose the View, Toolbars, Forms and select the spinner icon (sixth in the right-hand column). Move the cursor to the top left-hand side of E4, press the left mouse button and insert the spinner down to the lower center of E5. Release the mouse.
5 Select the spinner with the right mouse button and choose Format Control. Set the current value to 1, the minimum value to 0, the maximum value to 100, the incremental change to 1, and the cell link to E4 and click ok to close the spinner control dialog box.
6 Finally enter =E4/10 into D4 and center with red, bold, centered text and a single decimal place.
7 Copy this spinner (by right mouse button) into E6 with the same values but a cell reference of E6 and enter =E6/10 into D6 again in red, bold, centered text to one decimal place.
8 Copy a third spinner into E10 with a current value of 0, a minimum of 0, a maximum of 80, incremental change of 5, and cell reference of D10. Format D10 with red, bold text as a centered integer. This represents the model's location within the estuary.
9 Select the bathymetry sheet and enter 0 into C27, 5 into D27, 10 into E27, and continue until 80 is entered into S27. Center these cells.
10 Return to the main model, select B11, and enter

$$=\text{LOOKUP(D10,bathymetry!C27:S27,}$$

$$\text{bathymetry!C28:S28)}$$

to display the place name of the location being modeled, for the given names
11 Finally set S_2 to 1.0, M_2 to 2.0, and select station 30 which is the Bridge as shown in Figure 6.1.

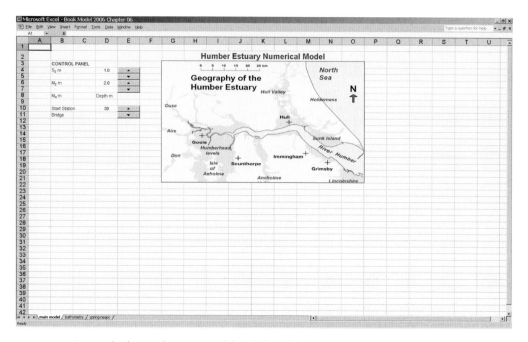

FIGURE 6.1 Screen display on the main model with the tidal controls completed.

6.4 MODELING SPRING-NEAP AMPLITUDES

The tidal model is set out to operate on single hour time steps in this section by entering two sets of text and one set of numbers:

1 Choose the spring-neaps sheet and select

B2	and enter	Time
C2		S_2 m
D2		M_2 m
E2		Total Tide m

2 Select B3 and enter 0, then B4 and enter =B3+1 with center alignment, and zero decimal places.

3 Select B4 to B339 and then fill down to count the hours for the full fourteen-day, spring-neap cycle.

4 Select C3 and enter

$$='main model'!\$D\$4*SIN(2*PI()*B3/12)$$

which is zero at $t = 0$ and can be read as the first term in Eq. 6.1: "main model"]!$D$4

is the amplitude of the S_2 species, SIN represents the sine function of, 2*PI()*B3 is the time in radians, and 12 is the period of the solar semi-diurnal tide in hours.

5 Select D3 and enter

$$='main model'!\$D\$6*SIN(2*PI()*B3/12.42)$$

which is zero at $t = 0$ and can be read as the second term in Eq. 6.1: "main model"!$D$6 is the amplitude of the M_2 species, SIN represents the sine function of, 2*PI()*B3 is the time in radians, and 12.42 is the period of the lunar semi-diurnal tide in hours.

6 Finally select E3 and enter

$$=C3 + D3 + 4$$

where the spring and neap S_2 and M_2 elevations are summed with the datum, which for the Humber is set at 4.0 m.

7 Select C3:E3, set to one decimal place, and then fill down to C339:F339 to calculate the tidal elevations throughout the spring-neap cycle as shown in Figure 6.2. Recall

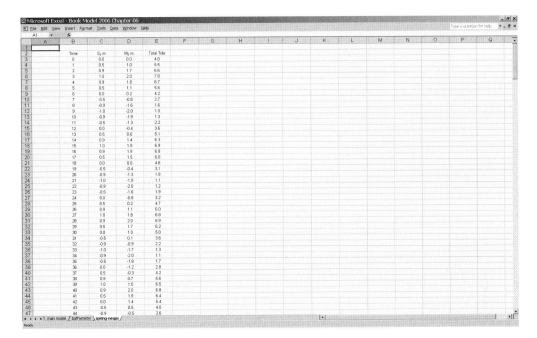

	Time	S_2 m	M_2 m	Total Tide
	0	0.0	0.0	4.0
	1	0.5	1.0	5.5
	2	0.9	1.7	6.6
	3	1.0	2.0	7.0
	4	0.9	1.8	6.7
	5	0.5	1.1	5.6
	6	0.0	0.2	4.2
	7	-0.5	-0.8	2.7
	8	-0.9	-1.6	1.6
	9	-1.0	-2.0	1.0
	10	-0.9	-1.9	1.3
	11	-0.5	-1.3	2.2
	12	0.0	-0.4	3.6
	13	0.5	0.6	5.1
	14	0.9	1.4	6.3
	15	1.0	1.9	6.9
	16	0.9	1.9	6.8
	17	0.5	1.5	6.0
	18	0.0	0.6	4.6
	19	-0.5	-0.4	3.1
	20	-0.9	-1.3	1.9
	21	-1.0	-1.9	1.1
	22	-0.9	-2.0	1.2
	23	-0.5	-1.6	1.9
	24	0.0	-0.8	3.2
	25	0.5	0.2	4.7
	26	0.9	1.1	6.0
	27	1.0	1.8	6.8
	28	0.9	2.0	6.9
	29	0.5	1.7	6.2
	30	0.0	1.0	5.0
	31	-0.5	0.1	3.6
	32	-0.9	-0.9	2.2
	33	-1.0	-1.7	1.3
	34	-0.9	-2.0	1.1
	35	-0.5	-1.8	1.7
	36	0.0	-1.2	2.8
	37	0.5	-0.3	4.2
	38	0.9	0.7	5.6
	39	1.0	1.5	6.5
	40	0.9	2.0	6.8
	41	0.5	1.9	6.4
	42	0.0	1.4	5.4
	43	-0.5	0.5	4.0
	44	-0.9	-0.5	2.6

FIGURE 6.2 Screen display showing the completed spring-neap modeling.

that S_2 is set to 1.0, M_2 to 2.0, and the station to 30 which is the Bridge.

6.5 MODELING M_4 AMPLITUDES

We noted earlier that the lunar quarter-diurnal is a harmonic of the lunar semi-diurnal generated by frictional effects in the estuary. The M_4 is added to the tidal model in three steps:

1 Select E8 on the main model sheet and enter

$$=LOOKUP(D10,bathymetry!C27:S27, bathymetry!C31:S31)$$

which displays the depth of the chosen station. Set to centralized alignment with one decimal place.

2 Select E9 on the main model sheet and enter

$$=LOOKUP(D10,bathymetry!C27:S27, bathymetry!C30:S30)$$

which displays the width of the chosen station. Set to centralized alignment with one decimal place.

3 Select C8 on the main model sheet and enter

$$=(3*1000*(80-D10)*D6*D6)/ (4*E8*6.21*3600*((9.81*E8)^0.5))$$

which is the amplitude of the M_4 from the third term in Eq. 6.1: $1000*(80-D10)$ is the distance upstream (m), D6*D6 is the square of the amplitude of the lunar semi-diurnal, E8 is the depth, 6.21*3600 and 9.81 are the period of the lunar semi-diurnal (s) and the gravitational acceleration (m s^{-2}), respectively. The result is 0.2 for the values given.

6.6 MODELING THE TIDAL WAVE

It is useful to set up a display of the tidal heights throughout the spring-neap cycle

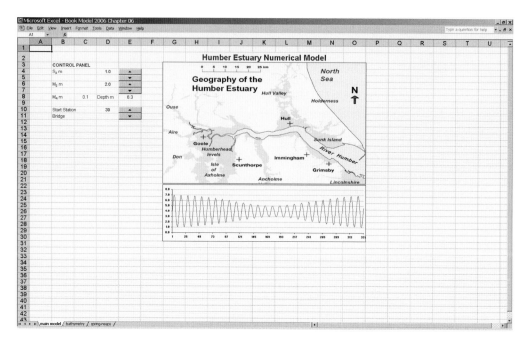

FIGURE 6.3 Plotting the spring-neap cycle on the main model spreadsheet.

within the estuary model. There are three simple steps:

1 Select E3:E339 on the spring-neap worksheet covering the water depths over the fourteen-day interval.
2 Choose Insert, Chart and follow the Wizard selecting line chart, smoothed line, and turning off the legend with "data are in columns" and then click to insert the chart onto the main model worksheet.
3 Pick up the chart and edit the axis so that it is located beneath the map G22:O30 on the main model display as shown in Figure 6.3.

6.7 GRAPHICAL DISPLAY OF THE SPRING-NEAP CYCLE

It is also useful to set up a display of the tidal heights throughout a single tidal cycle for the particular station within the estuarine model. There are six simple steps:

1 On the main model sheet

Select A31 and enter Time after
                              mid tide hrs
         A32   Lunar
                              quarter-diurnal m
         A33   Water depth m

2 Select D31 and enter 0. Select E31 and enter =D31 + 1. Center these cells and set to integer format. Then select E31 and fill right to R31.
3 Select D32 and enter

$$=\$C\$8*SIN(2*PI()*D31/6.21)$$

which is zero at $t = 0$ and can be read as the third term in Eq. 6.1: $\$C\$8$ is the amplitude of the M_4 species, $2*PI()*D31$ is the time in radians, and 6.21 is the period of the lunar quarter-diurnal tide in hours.

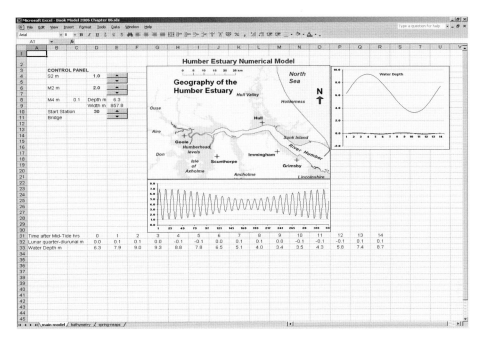

FIGURE 6.4 Plotting the water depths throughout the first tidal cycle on the main model spreadsheet.

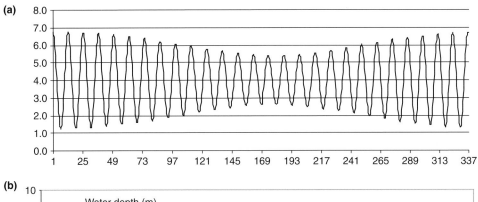

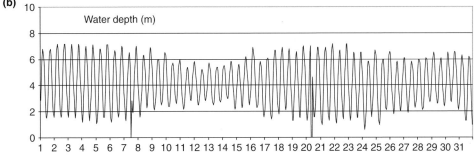

FIGURE 6.5 Comparison of (a) modeled spring-neap cycle with (b) Immingham tide gauge data for January 2003.

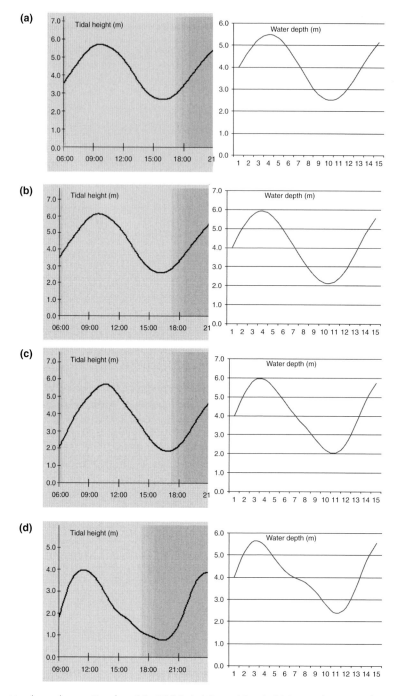

FIGURE 6.6 Humber tides on October 10, 2005: (a) Spurn Head, (b) Immingham, (c) The Bridge, and (d) Goole (left), and the modeled tides (right); with the S_2, M_2, and M_4 set as (a) 0.5, 1.0, and 0.0; (b) 0.6, 1.3, and 0.1; (c) 1.3, 0.6, and 0.3; and (d) 1.0, 0.4, and 0.5, respectively.

Center D32 and set to one decimal place.

4 Select D33 and enter

=LOOKUP(D31,'spring-neaps'!$B$3:
$B$339,'spring-neaps'!$E$3:
$E$339)+D32-4+$E$8

where LOOKUP determines the value corresponding to the time in D31 of the total tide on the spring-neap worksheet and adds the quarter-diurnal component from D32 and the cell's depth corrected for chart datum from E8. Center and set to one decimal place. The value should be 6.3 at $t = 0$ rising for example to 7.9 at $t = 1$.

5 Select D32:D33 and fill right to R32:33 to display the water depths throughout the tidal cycle.

6 Select D32:Q33 and insert smoothed line chart for the lunar quarter-diurnal and for the water depth including the S_2, M_2, and M_4 onto the main model spreadsheet as shown in Figure 6.4.

6.8 MODEL VALIDATION

The tides in the model estuary have many of the features of the natural tides described in the preceding chapters. Although based upon only three harmonics, the model tide rises and falls about twice every 24 hours, with ranges that vary throughout the spring-neap cycle. The model tide also becomes asymmetric as it travels upstream through the growth of the lunar quarter-diurnal.

For example, Chapter 5 gave A_{M2} and A_{S2} at Spurn Head as 2.13 and 0.72 m respectively which generates the model spring-neap cycle shown in Figure 6.5a and compares well with the observed cycle shown in Figure 6.5b for about January 6th to 20th, 2003 from the Immingham gauge. The period is necessarily correct and both model and field data range from HWST of about 6.8 m CD to HWNT of about 5.4 m. It will be noticed that there is a diurnal asymmetry present in the observed tides which is not included in the model. This is due to the fact that the solar orbital plane, the lunar orbital plane, and the Earth's equatorial plane are not co-incident.

The progress of the modeled tide along the estuary for individual tidal cycles is compared with the Humber's tides (UKHO Admiralty TotalTide, 2004) in Figure 6.6. It is apparent that, with judicious and justifiable choice of the species amplitudes, reasonable agreement is obtained. The asymmetry resulting from the relative growth of the lunar quarter harmonic is apparent. It is also apparent that the model uses a constant datum, whereas (Chapter 5) the datum falls from about 4.0 m at Spurn Head to about 2.2 m at Goole (Figure 6.6d).

Part IV

CURRENTS IN ESTUARIES

7

ESTUARINE CURRENTS

Contents

7.1 INTRODUCTION

This chapter provides an introduction to the physical principles which govern the flow of water and applies those principles to water flow in estuaries. Estuarine hydrodynamics are complex: the flood tide produces constantly changing water depths and transports thermal energy, solutes, and particulate matter in an upstream direction. The flood flow accelerates and decelerates to high water, before reversing as the ebb accelerates, decelerates, and empties the estuary. However, there are some simplifications with which generalized descriptions of estuarine currents can be derived and modeled. The simplifications are derived in this chapter and incorporated into the estuarine model in Chapter 8.

7.2 BACKGROUND INFORMATION

The chapter is based largely upon Hardisty (1990b) and Massey (1970) although there are a number of undergraduate texts which deal with the basic principles of fluid flow. Use has also been made of the more specialist oceanographic literature, in particular Neumann and Pierson (1966).

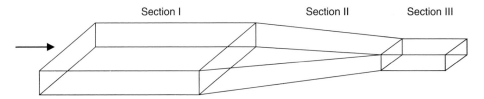

FIGURE 7.1 Schematic representation of the flow through a converging channel.

7.3 FLOW DESCRIPTORS

Consider the experiment shown in Figure 7.1 in which a flow channel with constant depth narrows in a downstream direction.

Three groups of terms are used to describe these flows. First, the flow in the wider, upper section is clearly slower than the flow in the narrower, lower section and the flow decelerates through the changing cross sections. From this, we define

1 **Uniform flow** has a constant velocity with distance (Sections I and III).
2 **Non-uniform flow** has a variable velocity with distance (Section II).

Second,

1 **Steady flow** is constant through time.
2 **Non-uniform flow** varies through time.

Third, a surface wave travels at a speed which depends upon the water depth (Chapter 5). The ability of a surface wave to travel upstream against the current thus depends upon the water depth. This provides another set of descriptive terms:

1 **Subcritical flow** exists when the surface wave is able to progress upstream against the current.
2 **Critical flow** exists when the surface wave travels upstream with the same celerity as the current.
3 **Supercritical flow** exists when the surface wave is unable to travel upstream against the current.

In general, estuarine flows accelerate, decelerate, and reverse over varying depths and are therefore unsteady, non-uniform, and can vary from subcritical, through critical, to supercritical within each tidal cycle.

7.4 THE REYNOLDS EXPERIMENT AND TURBULENCE

Materials, such as water, flow in response to applied stress. Different materials respond in different ways to the same amount of applied stress. If, for example, an eraser, a ball of clay, a cube of salt, and a cubic centimeter of honey are dropped onto the floor, each responds differently. The eraser rebounds, the soft clay sticks, the salt crystal fractures, and the honey spreads out slowly. These differences of behavior illustrate so-called elastic, plastic, fracture, and viscous deformation, respectively (Table 7.1). Elastic deformation is completely recoverable; viscous and plastic substances cannot reverse the deformation. Viscous substances such as honey and water respond to an applied stress by flowing and the rate of deformation (flow) depends on the magnitude of the stress. This is summarized by the fundamental relationship

$$\tau = \mu \frac{\mathrm{d}u}{\mathrm{d}z} \qquad \text{Eq. 7.1}$$

TABLE 7.1 Response of different substances to applied stress.

Material	Response	Deformation	Recovery
Eraser	Bounces	Elastic	Complete
Clay	Sticks	Plastic	Incomplete
Halite	Breaks	Fracture	None
Honey	Spreads out	Viscous	Flow

where τ is the shear stress, μ is the dynamic viscosity of water, and du/dz represents the velocity gradient. Materials which have such a direct relationship between the stress and the velocity gradient are known as Newtonian materials. Water is a typical Newtonian viscous fluid, and has a dynamic viscosity of about 0.01 poise (0.001 kg m^{-1}s^{-1}). Viscosity is thus the resistance to applied stress. The resistance to applied stress which occurs at the molecular level is known as the kinematic viscosity, v, and is equal to the dynamic viscosity divided by the mass density.

The behavior of water in an open estuarine channel is governed by the effects of the three forces, inertial, viscous, and gravitational, on the fluid. Much of the fluid's behavior depends upon the destabilizing effect of the inertial force compared to the stabilizing viscous force. In general, the inertial force is a function of the density of the fluid. The larger the density of a fluid, the greater is the force required to produce a specified acceleration in a specified volume.

This phenomenon was studied by Osborne Reynolds who carried out the experiment depicted in Figure 7.2 in the nineteenth century in Manchester, England. Reynolds noted that if the velocity is increased, the streak of dye in the water changed from a coherent and straight streak into a complex series of vortices and lateral mixings. The coherent flow represents laminar conditions where the stabilizing viscous forces dominate the inertial forces. The chaotic flow at higher speeds represents turbulent conditions where inertial instabilities dominate.

In laminar flow, a transfer of momentum from the faster to the slower water occurs at the molecular level whereas in turbulent flow whole packets of water are transferred. Eq. 7.1 applies directly to laminar flows, but for turbulent flows, it is modified to

$$\tau = K_z \frac{du}{dz} \qquad \text{Eq. 7.2}$$

where K_z is the coefficient of eddy viscosity which has a magnitude several orders larger than the molecular viscosity but the same dimensions.

7.5 THE REYNOLDS, FROUDE, AND RICHARDSON NUMBERS

The inertial force acting on a particle of fluid is equal in magnitude to the mass of the particle multiplied by its acceleration. The mass is equal to the density ρ multiplied by the volume (which is the cube of a length, L, characteristic of the geometry of the system). The mass is thus proportional to ρL^3. The acceleration of the particle is the rate at which its velocity changes with time and is thus proportional to the velocity divided by the time period. The time period may, however, be taken as proportional to the chosen characteristic length divided by the characteristic velocity so that the acceleration may

FIGURE 7.2   The Reynolds experiment (after Acheson (1990)).

be set proportional to $u/(L/u) = u^2/L$. The magnitude of the inertial force is thus

$$\text{Inertia Force} = \rho L^3 u^2/L = \rho L^2 u^2 \quad \text{Eq. 7.3}$$

The viscous force is given by Eq. 7.4 which is proportional to $\mu u/L$ where u is again a characteristic velocity and L is a characteristic length. The magnitude of the area over which this force acts is proportional to L^2 and thus the magnitude of the viscous force per unit area is

$$\text{Viscous Force} = (\mu u/L) \times L^2 = \mu uL$$
$$\text{Eq. 7.4}$$

Finally, the gravity force on the particle of water is equivalent to the particle weight, which is its mass multiplied by g, the gravitational acceleration. The magnitude of the gravity force is thus

$$\text{Gravity Force} = \rho L^3 g \qquad \text{Eq. 7.5}$$

There are thus three forces which control the stability or otherwise of the estuarine waters. The three ratios of a combination of two of these three forces representing the stabilizing to the destabilizing force, are expressed as three nondimensional numbers which characterize estuarine waters. The three nondimensional numbers are called the Reynolds, Froude, and Richardson numbers.

Reynolds number

The ratio of inertial forces to viscous forces determines whether the flow is laminar or turbulent, and is represented by the Reynolds number, R, which is the ratio of Eq. 7.3 and Eq. 7.4:

$$R = \frac{uL}{v} \qquad \text{Eq. 7.6}$$

where u is the velocity (ms^{-1}), L is a characteristic length, such as the depth of flow (m), and v is the kinematic viscosity $(m^2 s^{-1})$. The Reynolds number is dimensionless and can be a measure of any size of system and is independent of the units of measurement. For small values of R, viscous forces prevail and flow is laminar, but as R increases, turbulence sets in. The values of R which separate laminar and turbulent flow are called the transition zone and begin in the vicinity of Reynolds numbers of about 500–600. The upper boundary can be regarded as about 2000. In most open channels, laminar flow occurs rarely except near the boundary.

Froude number

The ratio of the destabilizing inertial forces to the stabilizing gravity force is also a measure of flow characteristics. The ratio is known as the Froude number, F, after William Froude (1810–1879), a pioneer in the study of naval architecture, who first introduced it. Nowadays the Froude number is usually taken as the square root of the ratio of the two forces given by combining Eq. 7.3 and Eq. 7.5:

$$F = \sqrt{\frac{\rho L^2 u^2}{\rho L^3 g}} = \frac{u}{\sqrt{gh}} \qquad \text{Eq. 7.7}$$

where the characteristic length scale, L, is replaced by the water depth h. The ratio of

flow velocity to the speed of movement of gravity waves is, thus, the Froude number (F) so that when $F > 1$, the flow is said to be super-critical; when $F < 1$, it is sub-critical. Critical flow conditions are represented by $F = 1$ as discussed in Section 7.3.

Richardson number

The ratio of the destabilizing gravity force to the stabilizing viscous force is known as the Richardson number R_I from Eq. 7.4 and Eq. 7.5:

$$R_I = \frac{\rho L^3 g}{\mu U L} \qquad \text{Eq. 7.8}$$

7.6 ESTUARINE MIXING PARAMETERS

A number of different schemes have been developed based upon these principles of fluid flow for the characterization of the complex flows which occur in estuaries (Figure 7.3).

Flow ratio, P

The simplest scheme was introduced by Simmons (1955) who noted that when the flow ratio (the ratio of river flow to the tidal prism) is 1.0 or greater, then the estuary is highly stratified, when the flow ratio is about 0.25, the estuary is partially mixed, and when it is less than 0.1 it is well mixed. Uncles et al. (1983) developed the idea for the Tamar Estuary by expanding the flow ratio, P, as

$$P = \frac{R}{AU_t} \qquad \text{Eq. 7.9}$$

where R is the river flow, A is the cross-sectional area, and U_t is the mean tidal current (so that AU_t approximates to

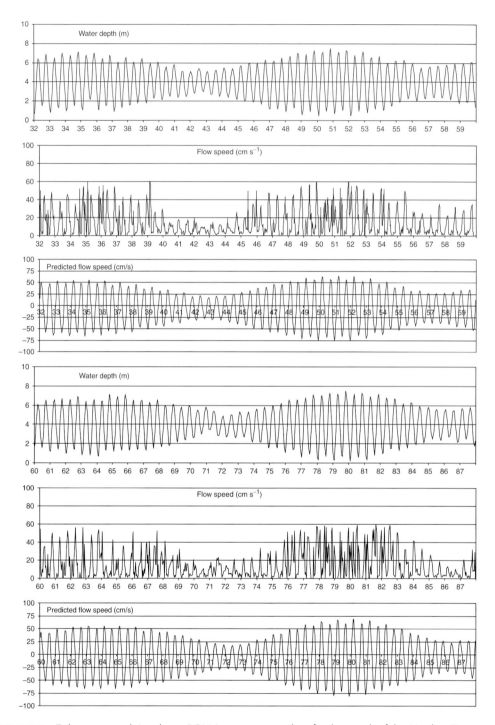

FIGURE 7.3 Tide gauge and Aanderaa RCM4 current meter data for the mouth of the Humber Estuary showing (top) current speed, depth, and resolved current speed for February 2004 and (bottom) for March 2004. Data from Hardisty (2004).

the tidal prism). The resulting values for P then vary from $P < 0.01$ for well-mixed to $P > 0.1$ for stratified conditions. However, Dyer (1997) notes that the Mersey and the Southampton water have flow ratios 0.01–0.02 and yet are partially mixed.

Estuary number, N_e

Dyer (1997) utilizes the flow ratio to define an estuary number, N_e:

$$N_e = \frac{PF_m^2}{TR} \qquad \text{Eq. 7.10}$$

where T is the tidal period and R is again the river flow as above. The F_m term was introduced by Hansen and Rattray (1966) and is called the densimetric estuary Froude number and is related to Eq. 7.7 as

$$F_m = \frac{u_f}{\sqrt{gh(\Delta\rho/\rho)}} \qquad \text{Eq. 7.11}$$

where u_f is the freshwater velocity, g is the gravitational acceleration, h is the water depth, $\Delta\rho$ is the ratio of the density difference between the seawater (which has a density ρ) and the freshwater. Dyer (1997) suggests that $N_e < 0.1$ corresponds to a well-mixed estuary and $N_e > 0.1$ indicates stratification.

7.7 STRATIFICATION NUMBER, S_t

The stratification number, S_t, is defined by Prandle (1985) as

$$S_t = \frac{0.85kU_oL}{(\Delta\rho/\rho)gh^2u_f} \qquad \text{Eq. 7.12}$$

where k is a friction coefficient (≈ 0.0025), L is the estuary length (m), U_o is the amplitude of the tidal currents, h is the water depth (m), and u_f is again the freshwater velocity. Prandle (1985) suggests that values of $S_t < 100$ indicate stratified conditions, $100 < S_t < 400$ indicate partially mixed conditions, and $S_t > 400$ correspond to well-mixed conditions.

Richardson number, R_I

The layered Richardson number (e.g. Mackay and Schumann, 1990; Kitheka, 2005) is defined as

$$R_I = \frac{gh(\rho_b - \rho_s)}{U^2\rho_0} \qquad \text{Eq. 7.13}$$

where U is the depth-averaged velocity (m s^{-1}), ρ_0 is the depth-averaged density (kg m^{-3}), h is the water depth (m), $(\rho_b - \rho_s)$ is the surface-bottom density difference (kg m^{-3}), and g is the acceleration due to gravity (9.81 m s^{-2}). Dyer and New (1986) observed that for $R_I < 2$ the bed-generated turbulence is the main mixing process and for $R_I > 20$ the water column is stable and bottom turbulence is not effective in mixing.

7.8 PROGRESSIVE AND STANDING TIDAL WAVES

The two end members in the continuum of surface wave characteristics are called progressive waves and standing waves. The tide in the deep ocean, far from land, has the characteristics of a progressive wave whereas by the time it achieves the upper reaches of an estuary and is reflected from the banks and bars it has the characteristics of a standing wave.

Progressive tidal wave

Flow velocities in a progressive wave are "in phase" with water depth as shown in Figure 7.4a in a process which is entirely analogous with waves breaking on a beach. Standing in the surf zone on a beach, we are all familiar with the fast, shoreward surge of water which accompanies the passage of the wave crest and this is followed by a weaker return flow which reaches a maximum as the wave trough passes. The tidal analogue, which is observed in deep water far from the shoreline, is a flood which reaches a maximum speed close to high water whilst the ebb reaches maximum speed close to low water.

Standing tidal wave

Within the confines of an estuary, however, the wave is reflected and evidences a behavior which is much closer to that of the bath tub. Consider a bath, half-filled with water. Consider that someone quickly jumps in and out of one end so that a wave is set up which reflects backwards and forwards about the mid-point. A current meter suitably deployed about one-quarter of the way from the tap end would, one

(a)

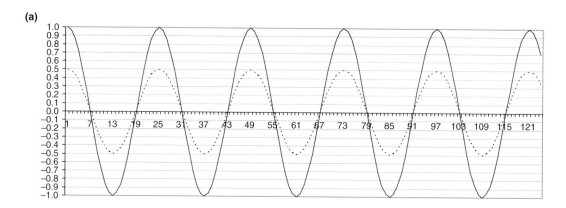

(b)

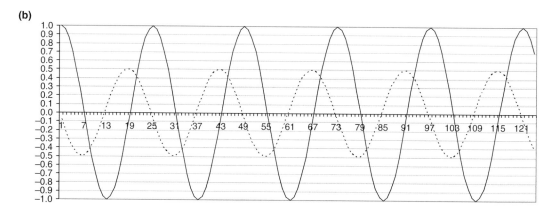

FIGURE 7.4 (a) Depth (solid line) and flow (broken line) relationships for a progressive tidal wave in deep water and (b) for a standing tidal wave in an estuary.

can imagine, record tap-ward flow as the water rushes toward the tap end, then slack water as the water attains maximum depth at the tap, and then the flow away which again reduces to zero when the wave has reached the far end. We say that the flow and water depths are out of phase as shown in Figure 7.4b. The estuary analog can also be imagined. In a small estuary at low water, the tide turns and the flood runs landward reaching a maximum at mid tide and then decreasing until the tide turns seaward at high water. The ebb reaches a maximum at mid tide but this time on a falling tide until the next low water is reached.

7.9 DISCHARGE RELATIONSHIPS

The flow discharge, $Q\,\mathrm{m\,s^{-3}}$, by mass continuity is simply the product of the flow speed $U\,\mathrm{m\,s^{-1}}$, and the cross-sectional area of the flow $A\,\mathrm{m^2}$ (Figure 7.5a):

$$Q = AU \qquad \text{Eq. 7.14}$$

Consider the two vessels in Figure 7.5b fill to volumes V_1 and V_2 respectively in a time t_f, with cross-sections of pipes 1 and 2 of A_1 and A_2 respectively. The actual flow speed can be calculated from

$$U_1 = \frac{V_1}{A_1 t_\mathrm{f}} \qquad \text{Eq. 7.15}$$

$$U_2 = \frac{V_1 + V_2}{A_2 t_\mathrm{f}} \qquad \text{Eq. 7.16}$$

where U_1 and U_2 are the flow speeds in pipes 1 and 2 respectively.

Eq. 7.15 and Eq. 7.16 can be developed to determine the tidal flow at any cross-section in an estuary $x = X$ as the product of the estuarine width, W_x, and the change in tidal depth per second, Δh_t, for the upstream length of the estuary divided by the cross-sectional area, $W_x D_x$. The freshwater contribution, u_f, is taken as a steady negative (out flowing) component equal to the freshwater discharge, Q, divided by the section's cross-sectional area:

$$u_\mathrm{f} = \frac{Q}{W_x D_x} \qquad \text{Eq. 7.17}$$

substituting for the exponential width and depth relationships introduced in Chapter 3.

$$U(x,t) = \frac{\int_{x=X}^{x=L} W_x \Delta h_\mathrm{t}\,\mathrm{d}x}{W_x D_x} - \frac{Q}{W_x D_x}$$

$$= \frac{\int_{x=X}^{x=L} W_0 e^{-ax/L} \Delta h_\mathrm{t}\,\mathrm{d}x - Q}{W_0 e^{-ax/L} D_0 e^{-bx/L}}$$

$$\text{Eq. 7.18}$$

7.10 SUMMARY

This chapter has introduced the principles of estuarine fluid flow including the concepts of unsteady critical flow and the Reynolds, Froude, and Richardson numbers. The tidal

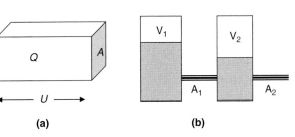

FIGURE 7.5 (a) Mass continuity determines the discharge volume through a cross-section in unit time as the product of the cross-sectional area and the flow velocity and (b) simple "jam jar" model of an estuary.

(a)　　　(b)

flow, $U(x, t)$ at time t and distance, $x = X$, from the mouth is given by

$$U(x, t) = \frac{\int_{x=X}^{x=L} W_0 e^{-ax/L} \Delta h_t dx - Q}{W_0 e^{-ax/L} D_0 e^{-bx/L}}$$

Eq. 7.19

where W_0 and D_0 are the estuarine width and depth at the mouth, L is the estuary's length, Δh_t is the change in tidal depth per second, Q is the freshwater discharge, and a and b are the width and depth coefficients, respectively.

8

MODELING CURRENTS

Contents

8.1 INTRODUCTION

The preceding chapters have demonstrated that tidal forcing generates currents in estuarine systems which are complex. The currents are inherently unsteady and will vary from subcritical to supercritical in each tidal cycle. However, the stage-discharge approach, which was detailed in Chapter 7, suggests that some simplifications are possible. In this chapter, the estuarine currents are modeled on the assumption that the changing water depths lead to known discharges through known cross-sections, so that the mean current (the ratio of the volume discharge to the cross-section) can be determined. The implementation of this model is described here, and tested by comparisons with the Humber Estuary.

8.2 BACKGROUND INFORMATION

It was shown in Chapter 7 after, for example, Dyer (1997, p. 37) that the freshwater and tidal discharge through an estuarine cross-section is the rate of change in water volume upstream from the cross-section. The rate of change of the water volume is, in turn, the sum of the product of the tidal depth change, the width and the length of the estuary upstream, and the freshwater contribution

$$U(x,t) = \frac{\int_{x=X}^{x=L} W_0 e^{-ax/L} \Delta h_t dx - Q}{W_0 e^{-ax/L} D_0 e^{-bx/L}}$$

Eq. 8.1

where W_0 and D_0 are the estuarine width and depth at the mouth, L is the estuary length, Δh_t is the change in tidal depth per second, Q is the freshwater discharge, and a and b are the width and depth coefficients, respectively.

Thus, the objective here is simply to calculate the volume of water, which passes through each cross-section and hence the mean tidal current throughout the tidal cycle.

8.3   MODELING UPSTREAM VOLUME CHANGES

The estuarine bathymetry is used to calculate the volume of water discharged through the cross-section. There are four stages in this work:

1 Open the model and save as with a new version number, then select C33 on the bathymetry worksheet and enter:

$$=5000*C30/1000000$$

where the formula is calculating the volume of water contained per meter of tidal depth in this cell:

5000	is the length of the cell (m)
C30	is the width of the cell (m)
/1000000	translates the result into 10^6 m³. The result is displayed at this stage as 0.880755.

2 Select D33 and enter:

$$=5000*D30/1000000 + C33$$

where the formula is calculating the volume of water contained per meter of tidal depth in this cell plus all other

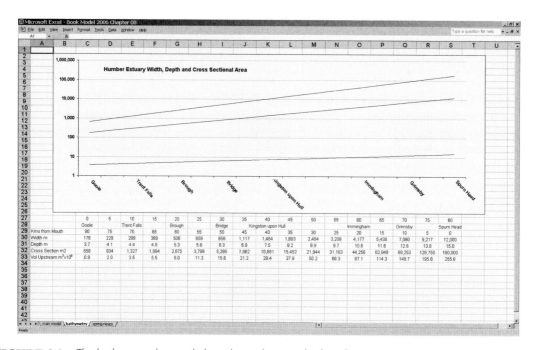

FIGURE 8.1   The bathymetry sheet with the volume changes displayed.

upstream cells (C33). The result at this stage is 2.021749.

3 Select C33:D33 and format to one decimal place and center alignment.

4 Select D33 and fill right to S33 to display the volume of water contained per meter of tidal depth upstream of the downstream boundary of each of the sixteen cells in the model estuary. For the whole estuary this is $255.8 \times 10^3 \, \text{m}^3$ (Figure 8.1).

8.4   MODELING THE TIDAL FLOW

In this section the mean tidal currents through the cycle are calculated by dividing the discharge by the cross-sectional area:

1 Choose the main model sheet and select

S33   and enter   At this station

R34   Cross-section

T34   =LOOKUP(D10,bathymetry!C27:S27, bathymetry!C32:S32) and set to integer centered text with a comma thousands.

U34   and enter   m²

R35   and enter   Upstream vol/m tide

T35   =LOOKUP($D$10,bathymetry! C27:S27,bathymetry!C33:S33) and set to centered text with one decimal place.

U35   and enter $\text{m}^3 \times 10^6$

2 The model displays, for example, a cross-section as 5,395 m² and the upstream volume change per meter of tide as $15.6 \, \text{m}^3 \times 10^6$ for the Bridge station.

3 Enter Tidal current m s⁻¹ into cell A34.

4 Select D34 and enter

$$= \$T\$35 * 1000000 * (E33-D33)/$$

$$(\$T\$34 * 3600)$$

which can be read as Eq. 8.1:

$T$35 is the discharge through the cross-section for each meter change in depth;

E33-D33 is the actual change in tidal water depth each hour;

$T$34 is the cross-section at the chosen station;

3600 converts the flow from meter per hour to meter per second.

5 Select D34 and set to one decimal place with centered text. 1.3 is displayed.

6 Finally select D34:Q34 and then fill right to calculate the tidal currents throughout the tidal cycle.

With the tidal inputs specified in Chapter 6 ($S_2 = 1.0$, $M_2 = 2.0$) and for the Bridge Station this gives a tidal current from 1.3 at $t = 0$ to -1.0 at $t = 6$ (Figure 8.2).

8.5   MODELING THE FRESHWATER FLOW

The freshwater flow can be modeled by dividing the input by the cross-section at the chosen station and displaying the result:

1 Select   B12   and enter   Freshwater input

F12   cumecs

B13   Flow at station

F13   m s⁻¹

2 Copy a spinner into E12 and set this with a minimum current value and increment of 10 and a maximum of 2,000 referenced to cell D12.

3 Center align D12 with integer format in bold and red.

4 Select D13 and enter =D12/T34. Center with two decimal places to display the mean freshwater current at the chosen station.

The result for a freshwater input of 200 cumecs at station 10 is a freshwater current of 0.15 m s⁻¹ as shown in Figure 8.3.

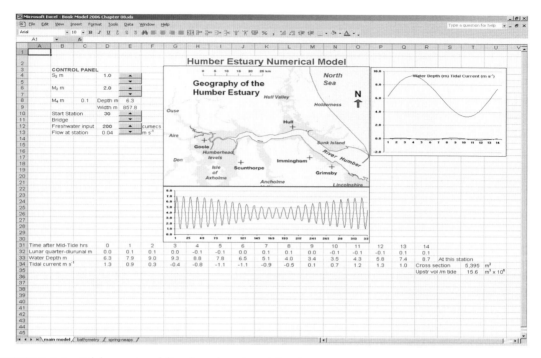

FIGURE 8.2 Tidal current modeling throughout the tidal cycle.

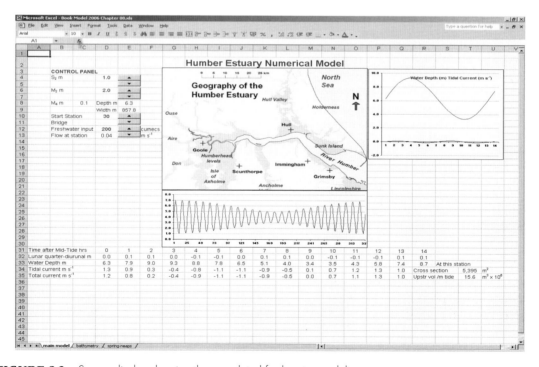

FIGURE 8.3 Screen display showing the completed freshwater modeling.

(a)

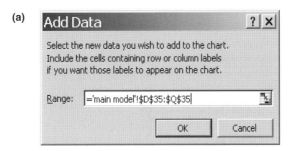

(b)

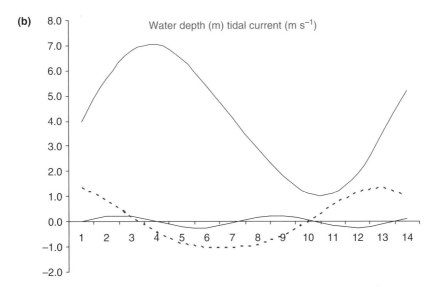

FIGURE 8.4 (a) Adding the flow data to the chart and (b) the details of the water depth (solid lines) and flow (broken line).

8.6 MODELING THE TOTAL FLOW

The total current can be modeled by simply subtracting the outgoing (and therefore negative) freshwater flow from the tidal component:

1 Select A35 and enter Total Current m s^{-1}.
2 Select D35 and enter =D34−$D$13.
3 Set D35 to center alignment and one decimal place, then fill right to Q35.

The result for the inputs above is a flood (positive) flow of 1.3 m s^{-1} at $t = 0$, and ebb (negative) of 1.1 m s^{-1} at $t = 6$ returning to a maximum flood of 1.4 m s^{-1} at $t = 12$.

8.7 GRAPHICAL DISPLAY OF THE FLOW

It is useful to set up a display of the tidal currents and to compare these with the surface water depths within the estuary model:

1 Return to the Bridge station with a freshwater input of 200 cumecs and S_2 and M_2 amplitudes of 1.0 and 2.0 m respectively.
2 Select the water depth chart and then Add Data from the Chart Menu typing ='main

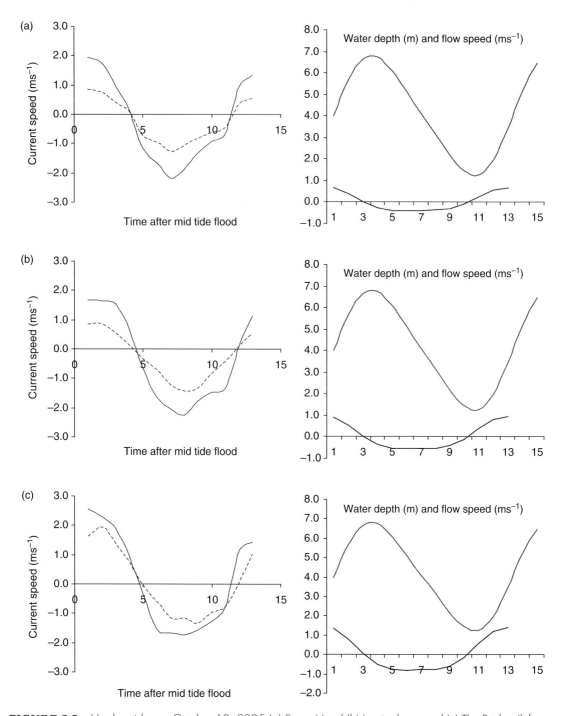

FIGURE 8.5 Humber tides on October 10, 2005 (a) Spurn Head (b) Immingham, and (c) The Bridge (left, solid lines springs and broken lines neaps) and the modeled tides (right) with the S_2, M_2, and M_4 set as 0.9, 1.8, and 0.4 respectively.

model'!$D$35:$Q$35 into the Range box as shown in Figure 8.4(a).

3 Adjust the title and the axes as shown in Figure 8.4(b) to display the currents beneath the water depths throughout the tidal cycle. Note that flood is positive and ebb is negative in the model.

8.8 MODEL VALIDATION

It is apparent that the currents in the model estuary have many of the features of the natural flows described in the preceding chapter. Although based upon only three harmonics and the mass continuity approach, the model tide floods and ebbs twice every 24 h. The tide achieves quite realistic values of $1-2\,m\,s^{-1}$ which increase when spring tide ranges are input in comparison with neap values. The model tide can also become asymmetric as it travels upstream through the growth of the M_4 as discussed in Chapter 6. The residual or freshwater flow also becomes vanishingly small in the model as in reality when the mouth of the estuary is approached because of the non-linear increase in cross-section.

In particular Figure 8.5 compares the charted tidal currents with the modeling results for a neap tide. Spring maxima are about $1\,m\,s^{-1}$ at Spurn and Immingham rising toward $2\,m\,s^{-1}$ at the Bridge and are reasonably modeled although there is a pronounced ebb asymmetry in the data which is not reproduced in the model.

Part V

THE TEMPERATURE AND SALINITY OF ESTUARIES

9

ESTUARINE TEMPERATURE AND SALINITY

Contents

9.1 INTRODUCTION

Water properties are generally taken to include all of the organic and inorganic substances transported in solution or suspension together with a range of physical attributes of the water itself. Most of these properties vary both geographically throughout the estuarine system, and temporally on tidal, intertidal, or seasonal time scales. In this chapter we cover perhaps the two most important properties: the physical temperature of the water and the chemical salinity due to the dissolved salts.

9.2 BACKGROUND INFORMATION

Water consists of covalently bonded oxygen and hydrogen atoms with the single electron from each of two hydrogen atoms sharing one of the six electrons in the outer, third shell of the oxygen atom so that completed shells are formed on all three atoms (as shown in Figure 9.1). The water molecule is therefore H_2O. The chemical composition of pure water does not, however, offer any explanation for the great dissolving power of the liquid. The dissolving power is a direct consequence of the ability of the water

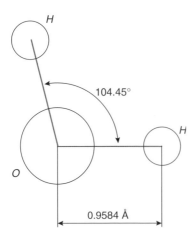

FIGURE 9.1   Molecular structure of pure water.

molecules to disassociate solute ions, and is due to water's uniquely high dielectric constant.

The dielectric constant, ε_e, of any material is a number which expresses how much smaller the electric intensity is in a space filled by the material than in a vacuum if the same electric field is applied. This ratio is expressed by considering a condenser with capacitance C_o in a vacuum which changes to C_m with a particular material between the plates. Then,

$$\varepsilon_e = \frac{C_m}{C_o} \qquad \text{Eq. 9.1}$$

ε_e has a value of 1 for a vacuum by definition, but this increases to a value of 1.0006 for air, 2.0 for petroleum, 5–7 for glass, 6–8 for mica, and up to 81 for water. This is the highest dielectric value of all liquids. The reason for water's large ε_e value lies in a fortuitous peculiarity in the structure of the H_2O molecule. The peculiarity of the water molecule is that the two valence electrons of oxygen join the valence electrons of the hydrogen atom not at 180° as might intuitively be thought, due to the mutual repulsion of the hydrogen atoms, but at an angle of 104°. The result is a molecule with a positively charged hydrogen region

and a negatively charged oxygen region, which results in a so-called dipole moment within the molecule. This, in turn, accounts for two of water's fundamental properties.

First, Coulomb's Law states that the force of the electrical attraction between two charges is inversely proportional to the dielectric constant of the surrounding media. Thus the force of attraction between the positively charged sodium atom and the negatively charged chlorine atom in common salt, NaCl, is considerably reduced when in water because of the liquid's very high dielectric constant. That is why water is a very powerful solvent, and the sea saline. Second, the dipole moment is the reason for the existence of strong forces acting between the water molecules themselves. The negatively charged hydrogen regions of one molecule attract the positively charged oxygen regions of another so that two, three, or more H_2O molecules are bonded together and form multiple groups known as polymers depending largely on the temperature of the liquid. At a temperature of 0°C the groups correspond to $(H_2O)_6$, but the degree of polymerization decreases with increasing temperature. The physical properties of water, including its high surface tension, high melting and boiling points, and its peculiar change in density with temperature, are all due to this polymerizing behavior, and thereby to the dipolar nature of the molecule.

9.3   TEMPERATURE

The world ocean has a surface area of some 360 million km^2, made up of a series of large interconnected basins and smaller continental seas. In order to understand the distribution of thermal energy in estuaries, it is necessary to begin by considering the source

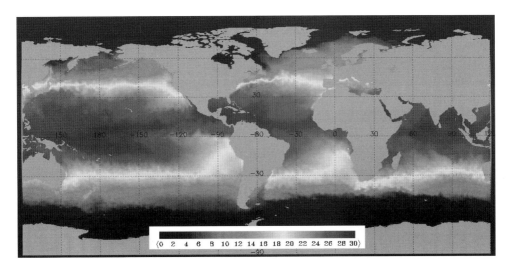

FIGURE 9.2 Global distribution of ocean surface temperatures (DIR 9.1).

of heat for the world's oceans as a whole. There is a constant inflow of energy from the Sun and a constant outflow of radiation from the Earth back to space. The largest thermal energy source is heat absorbed from solar and sky radiation. At the upper limit of the Earth's atmosphere this produces on average an input of some $700 \, \text{gcal} \, \text{cm}^{-2} \, \text{day}^{-1}$. About 57% is absorbed by the atmosphere and the balance reaches the sea surface.

The sea temperature at the mouths of estuaries thus varies geographically (Figure 9.2) and also on a seasonal basis, although sea temperatures are generally more stable than the temperatures of the freshwater input.

9.4 SALINITY

The relatively high salt content of seawater is of special importance in oceanography and is quantified through the measurement of salinity, which is defined as the weight of dissolved solids in gram per 1000 g of water, usually expressed in parts per thousand (ppt) and symbolized by ‰.

In the open ocean the salinity varies from about 34‰ to about 38‰, with an average close to 35‰. Landlocked seas in humid regions with a large freshwater river output and little evaporation often have a much lower salinity; values of less than 1‰ are reported from open fjords and from the Baltic Sea. Conversely, landlocked seas in arid regions, where evaporation concentrates the brine, can have much higher values, with more than 45‰ being recorded in the Red Sea and the Dead Sea.

Careful chemical titrations were traditionally undertaken to determine the salinity and the chemical composition of seawater. Work since Dittmar's nineteenth-century analysis of samples from the world's oceans collected on the famous Challenger Expedition has shown that, regardless of the absolute concentration, the relative proportions of the different salts in seawater are constant to the second decimal place of salinity. The proportions are shown in Table 9.1.

TABLE 9.1 Major constituents of seawater.

Cations			Anions		
Sodium	Na^+	30.62	Chloride	Cl^-	55.07
Magnesium	Mg^{++}	3.68	Sulphate	$SO4^{--}$	7.72
Calcium	Ca^{++}	1.18	Bicarbonate	HCO_3^-	0.40
Potassium	K^+	1.10	Bromide	Br^-	0.19
Strontium	Sr^{++}	0.02	Borate	$H_2BO_3^-$	0.01

Figures are percentages by weight of the total major constituents.

In 1978 the break with chlorinity was sealed with the introduction of a new conductivity-based definition by the Joint Panel on Oceanographic Tables and Standards (JPOTS). This new definition, which is current today, states that "a seawater of salinity 35 has a conductivity ratio of unity with 32.4356 grams of Potassium Chloride in 1 kilogram of solution at 15°C and 1 atmosphere." The standard concentration of KCl was derived from measurements carried out on one batch of standard seawater and measurements of absolute conductivity carried out at the Institute of Oceanographic Sciences, Wormley, UK.

The introduction of the Practical Salinity Scale 1978 (PSS78) resulted in some significant operational changes. Previous methods had resulted in concentration units such as parts per thousand (ppt) and symbols such as ‰ being used but these were no longer valid. The correct way to report practical salinity is as a number (e.g. the sample had a salinity of 35) with reference to PSS78. However the move away from units has proved so difficult for some to accept that a new unit, the Practical Salinity Unit (PSU), has been unofficially introduced. In true terms of the PSS78 the PSU is not valid but it is widely used and often not rejected by journal editors.

9.5 ADVECTION AND DIFFUSION

Estuarine waters respond slowly to diurnal solar heating effects and conduct heat and transfer salt only slowly. Most changes are due to advection which is the movement of water masses that transport warm water to cooler regions or saltier water to fresher regions and *vice versa*. However, the longer term values depend upon diffusion due to turbulent mixing (Chapter 7). The turbulent flow components at any instant are represented as,

$$u = \overline{u} + u' \qquad \text{Eq. 9.2}$$

$$v = \overline{v} + v' \qquad \text{Eq. 9.3}$$

$$w = \overline{w} + w' \qquad \text{Eq. 9.4}$$

where $\overline{u}$ is the mean component in the x direction (normally downstream), $\overline{v}$ is the mean component in the y direction, $\overline{w}$ is the mean component in the z direction (vertical), and u', v', and w' are the fluctuations about the respective means.

The statistical approach studies the histories of the motion of individual fluid parcels and attempts to determine from these the statistical properties necessary to represent diffusion, rather than considering the material or momentum flux at a fixed space point. Perhaps the simplest model of diffusion is the well-known random walk which is essentially a simple discrete step stochastic diffusion model. Each step in the random walk is independent of previous motion, and in the analogous diffusion model the movement of a pollutant is uncorrelated with its previous motion. The molecular analog of this model describes Brownian diffusion and for large time intervals the distribution approaches the normal or Gaussian distribution.

9.6 THE GAUSSIAN DISTRIBUTION

Models which predict the value of parameters at some stage in the diffusing process rest on the assumption that, although the volume occupied by the material increases with time, the total mass remains constant (Lewis, 1997). If the mass is known the problem reduces to defining the size of the volume after some specified time of diffusion. The distribution in a given direction frequently has a bell-shaped form, which may resemble a Gaussian function, and the spread of the distribution can be expressed by its variance. It is usual to describe this variance, σ_x, of a concentration distribution as follows:

$$\sigma_x^2 = \frac{\int_{-\infty}^{\infty} C(x) x^2 \, dx}{\int_{-\infty}^{\infty} C(x) \, dx} \qquad \text{Eq. 9.5}$$

where $C(x)$ is the concentration at any position x. In this expression the denominator represents the total mass of the substance in a section. When the distribution is truly Gaussian, the function describing the variation is as follows:

$$C(x) = \frac{1}{\sqrt{2\pi}\,\sigma_x} \exp\left[-\frac{x^2}{2\sigma_x^2}\right] \qquad \text{Eq. 9.6}$$

This expression can be normalized to produce a distribution about unity by simply removing the multiplier to give (Figure 9.3)

$$C(x) = \exp\left[-\frac{x^2}{2\sigma_x^2}\right] \qquad \text{Eq. 9.7}$$

This distribution can be fitted to a diffusing dye or pollution patch and to the diffusion of thermal energy and temperature in estuarine environments.

9.7 ESTUARINE TEMPERATURES

Water temperature in estuarine systems varies throughout the tidal cycle at any sampling location if there is a difference between the temperature of the seawater and the freshwater flow. In general, two cases are usually recognized:

Intratidal temperature winter case

In winter, the seawater temperature is usually warmer than the freshwater, so that water temperature at a sampling station will rise with the flooding of the tide and reach a maximum at high water (Figure 9.4).

Intratidal temperature summer case

In summer, the seawater temperature is generally cooler than the freshwater, so that water temperature at a sampling station will fall with the flooding tide and reach a minimum at low water.

The temperature will also vary longitudinally at any instant during the tidal cycle if there is a difference between the temperature of the seawater and

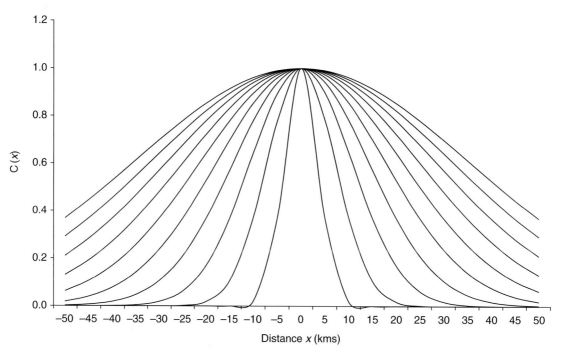

FIGURE 9.3 Normalized Gaussian distribution from Eq. 9.7 with a σ_x value from 2 to 20 (inner to outer plots) and x (km) (Toolbox 6).

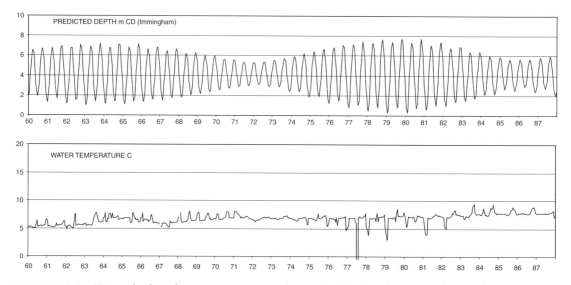

FIGURE 9.4 Water depth and water temperature at the mouth of the Humber Estuary for March 2004. Data from Hardisty (2004).

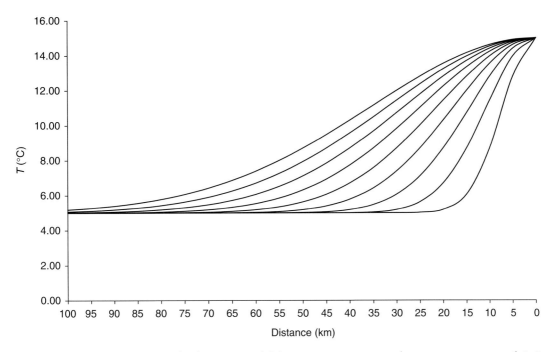

FIGURE 9.5 Using a Gaussian distribution to model the estuarine temperature for a river temperature of 5°C and a sea temperature of 15°C from Eq. 9.8 for σ_x from 4 (lower line) to 20 (upper line) (Toolbox 7).

the freshwater flow. In general, two cases are again recognized.

Longitudinal temperature winter case

In winter, the seawater temperature, T_S°C, is usually warmer than the river, T_R°C, so that water temperature decreases in an upstream direction. Eq. 9.7 can be adopted to describe this distribution as a Gaussian curve:

$$T(x) = (T_S - T_R)\exp\left[-\frac{x^2}{2\sigma_x^2}\right] + T_R$$

$$\text{Eq. 9.8}$$

as shown in Figure 9.5 for comparison with Gameson's (1982) data for the Humber Estuary. The data appear closely to fit the curve with a value of 10 and 20°C for the river and sea temperatures, and a variance of 24,000 for distance in kilometers.

Longitudinal temperature summer case

In summer the seawater temperature is generally cooler than the freshwater, so that water temperature increases in an upstream direction. Eq. 9.7 can again be adopted to describe this distribution as a Gaussian curve.

$$T(x) = (T_R - T_S)\exp\left[-\frac{x^2}{2\sigma_x^2}\right] + T_S \quad \text{Eq. 9.9}$$

9.8 ESTUARINE SALINITIES

The salinity in an estuarine system will vary throughout the tidal cycle, generally rising toward high water and falling with the ebb tide. The situation may be more complicated

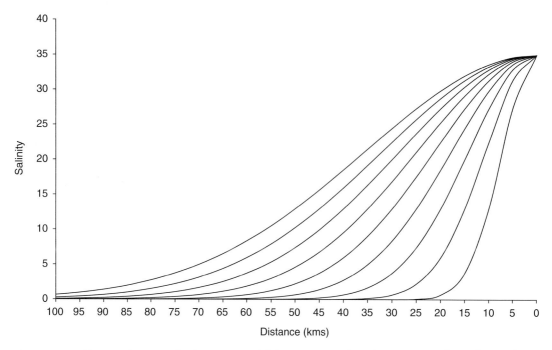

FIGURE 9.6 Using a Gaussian distribution to model the estuarine salinity from Eq. 9.11 for σ_x from 4 (lower line) to 20 (upper line) (Toolbox 8).

in partially mixed or salt wedge estuaries; this discussion will however concentrate on well-mixed examples. Additionally, there is a net downstream flux of salt due to the freshwater flow which is counterbalanced by the upstream diffusion due to turbulence and mixing. Many authors (e.g. Dyer, 1986) present detailed analyses of this longitudinal distribution in terms of the salt balance which can be expressed as

$$\bar{u}\frac{\delta\bar{s}}{\delta x} = \frac{\delta}{\delta x}\left(K_S\frac{\delta\bar{s}}{\delta x}\right)$$ Eq. 9.10

This states that the longitudinal advection of salt downstream on the sectional mean flow (the left-hand side of Eq. 9.10) is balanced by an upstream diffusion (the right-hand side of Eq. 9.10). This approach has been used by, for example, West and Williams (1975) in the Tay Estuary in Scotland. The seawater salinity can usually be assumed to be 35‰ whereas the river water can be assumed to be fresh, so that salinity decreases in an upstream direction. Eq. 9.7 can be adopted to describe this distribution as a Gaussian curve:

$$S(x) = 35\exp\left[-\frac{x^2}{2\sigma_x^2}\right]$$ Eq. 9.11

where an example is shown in Figure 9.6 for comparison with Gameson's (1982) data for the Humber Estuary. The data appear closely to fit the curve with a variance of 24,000 for distance in kilometers.

9.9 SUMMARY

This chapter has introduced the concepts of temperature and salinity in estuarine environments and described tidal and longitudinal variations in both parameters. It has also explained advection and diffusion and suggested that Gaussian distributions may be used to describe the longitudinal variation in temperature (Eq. 9.12 and Eq. 9.13) and salinity (Eq. 9.14):

$$T(x) = (T_S - T_R) \exp\left[-\frac{x^2}{2\sigma_x^2}\right] + T_R$$

$$\text{Eq. 9.12}$$

$$T(x) = (T_R - T_S) \exp\left[-\frac{x^2}{2\sigma_x^2}\right] + T_S$$

$$\text{Eq. 9.13}$$

$$S(x) = 35 \exp\left[-\frac{x^2}{2\sigma_x^2}\right] \qquad \text{Eq. 9.14}$$

where the seawater temperature $T_S°C$, is warmer than the river $T_R°C$, in Eq. 9.12 and *vice versa* in Eq. 9.13; x is distance from the mouth, and σ_x is a dispersion coefficient.

10

MODELING TEMPERATURE AND SALINITY

Contents

10.1 INTRODUCTION

Chapter 9 covered the advection and diffusion of thermal energy and salt in estuarine environments. The use of the Gaussian approach was discussed and typical cases were examined. In this chapter, the Gaussian distribution is used to represent the diffusive distribution of temperature and salinity in the estuarine environment in the model. Algorithms are then incorporated that advect the distributions through the tidal cycle. The resulting temperature and salinity distributions, and the differences between, for example, summer and winter temperatures are examined.

10.2 BACKGROUND INFORMATION

The distribution of a parameter in a given direction frequently has a bell-shaped form, which may resemble a Gaussian function (Lewis, 1997), and the spread of the distribution can be expressed by its variance. If the distribution is Gaussian, the function $P(x)$ describing the variation of property P in the x direction is given by

$$P(x) = K_P \frac{1}{\sqrt{2\pi}\,\sigma_P} \exp\left[-\frac{x^2}{2\sigma_P^2}\right] \quad \text{Eq. 10.1}$$

where K_P is an appropriate coefficient and σ_P is the variance. For a normalized curve in the Humber Estuary, with distances in kilometers, σ_P is about 24 (Chapter 9) and hence Eq. 10.1 becomes

$$P(x) = \exp\left[-\frac{x^2}{2\sigma_P^2}\right] \qquad \text{Eq. 10.2}$$

This expression is utilized to model temperature $T(x)$ and salinity $S(x)$ at distance x from the mouth as

$$T(x) = (T_S - T_R)\exp\left[-\frac{x^2}{2\sigma_x^2}\right] + T_R$$
$$\text{Eq. 10.3}$$

$$T(x) = (T_R - T_S)\exp\left[-\frac{x^2}{2\sigma_x^2}\right] + T_S$$
$$\text{Eq. 10.4}$$

$$S(x) = 35\exp\left[-\frac{x^2}{2\sigma_x^2}\right] \qquad \text{Eq. 10.5}$$

where the seawater temperature $T_S\,°C$ is warmer than the river $T_R\,°C$, in Eq. 10.3 and *vice versa* in Eq. 10.4; σ_x is a dispersion coefficient.

10.3 MODELING A GAUSSIAN PROCESS

In this section we introduce a normalized Gaussian distribution, which then can be calibrated for temperature and salinity to model the respective distributions along the estuary. There are six stages in this work:

1 Open the model developed from Chapter 8 and "save as" with a different version number.
2 On the "bathymetry" sheet select

A35	and enter	Gaussian
A36		Reverse Gaussian
T27		80

3 Select C35 and enter $=2.7^{\wedge}((-C29*C29)/1152)$ where 2.7 is the value of the

exponential in Eq. 10.2 and C29 is the distance in kilometers from the mouth of the estuary.
4 Center align C35 with two decimal places then fill right to T35.
5 Select C36 and enter $=2.7^{\wedge}((-C27*C27)/1152)$ where the formula is the same as above, except that the distances are measured from the head of the estuary.
6 Select C36 and center align with two decimal places then fill right to T36 to model a reverse Gaussian distribution along the estuary as shown in Figure 10.1.

10.4 THE TEMPERATURE DISTRIBUTION

The Gaussian distribution is used to model temperature variations along the estuary:

1 On the main model spreadsheet:

select	B14	and enter	Temp °C
	C15		River
	E15		Sea

2 Copy a spinner into D14:D15 and F14:F15.
3 Set the D14 spinner to a current value, minimum and increment of 1, a minimum of 1, a maximum of 30, and a cell reference of C14.
4 Repeat this process with the F14 spinner with identical settings but a cell reference of E14.
5 Select C14 and E14 and format to a single decimal place, centered text in red bold, as shown in Figure 10.2 with a river temperature of 5 and a sea temperature of 15.
6 Select the bathymetry spreadsheet and

Select	A37	and enter	Temperature °C
	A2		River T
	A3		="main model"!C14
	A4		Sea T
	A5		="main model"!E14

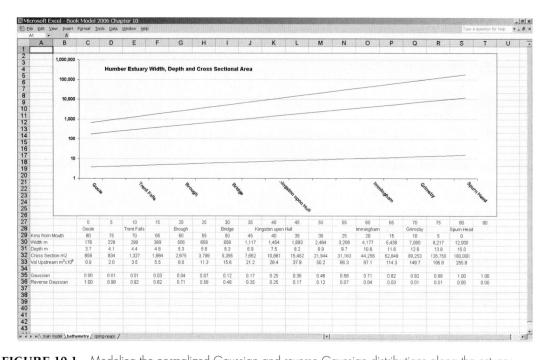

FIGURE 10.1   Modeling the normalized Gaussian and reverse Gaussian distributions along the estuary.

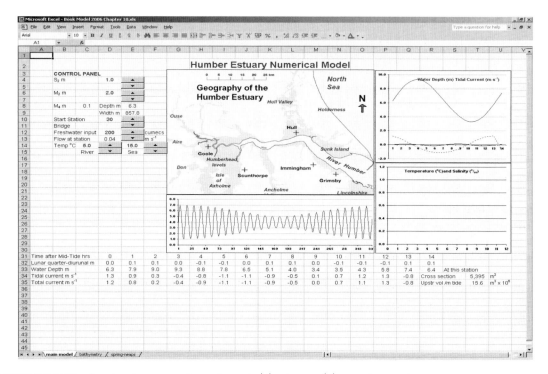

FIGURE 10.2   Inserting two new spinners to control the river and the sea temperatures.

7 Select C37 and enter

$$= IF(\$A\$5>\$A\$3, \$A\$3+(\$A\$5-\$A\$3)$$
$$*C35, \$A\$5+(\$A\$3-\$A\$5)*C36)$$

where the conditional if statement utilizes the normalized Gaussian if the sea temperature is greater than the river temperature (A5 > A3), and otherwise utilizes the normalized reverse Gaussian.

8 Select C37 and center with two decimal places then fill right to S37 to model the temperature distributions along the estuary as shown in Figure 10.2. For example, if the river temperature is set to 5 and the sea temperature is set to 15, then the estuarine temperatures increase from 5.04 at Goole, through 7.52 at Kingston upon Hull to 15.00 at the mouth.

It is now necessary to advect the thermal energy distribution landward and seaward along the estuary with the flooding and ebbing of the tide. It is assumed in this model that temperature distribution defined on the bathymetry spreadsheet represents conditions at mid tide on the flood. There are six stages in this work:

1 Select cell A36 on the main model spreadsheet and enter Displacement kms.
2 Select D36 and enter =C36+(3600*D35)/1000. to calculate the total distance advected from $t = 0$ in kms.
3 Format D36 to one decimal place centered and then fill right to Q36. The water is displaced 9.7 km downstream (negative) with the earlier tidal settings before turning at low water and then returning upstream.
4 Select A37 and enter Temperature °C
5 Select D37 and enter

$$= LOOKUP(\$D\$10+D36,bathymetry!$$
$$\$C\$27:\$S\$27, bathymetry!\$C\$37:\$S\$37)$$

6 Select D37 and format to two decimal places and fill right to P37.

The result, for the settings given earlier, is that the temperature at the Bridge rises from 6.16 to 6.74 as the warmer seawater floods upstream and then falls to 5.45 as the colder river water is advected downstream by the ebbing tide (Figure 10.3).

10.5 Displaying the temperature distribution

The temperature distribution is displayed as both a longitudinal profile on the bathymetry sheet and as a new time series on the main model sheet:

1 Select the bathymetry sheet and click on the chart and then select the add data command in the chart dialog box.
2 Select the temperature data cells C37 to S37 and click ok.
3 Select the new line on the chart and format it appropriately as shown in Figure 10.4.
4 Select D31:P31 and D37:P37 on the main model spreadsheet by pressing the left mouse button to select the first array, then Ctrl before releasing the mouse button and selecting the second array before releasing Ctrl.
5 Insert an x–y line chart from the menu bar and follow the wizard to insert the chart onto the main model spreadsheet as shown in Figure 10.5.

10.6 The salinity distribution

Here, the Gaussian distribution is again used to model salinity variations. This is rather simpler than modeling temperatures because the salinity of the seawater will be assumed to be constant at 35‰ and the river

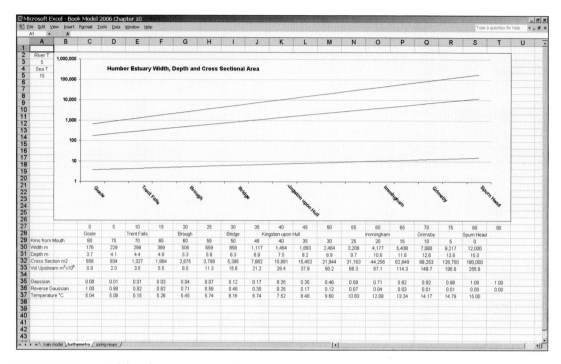

FIGURE 10.3 Modeling the temperature variations along the estuary.

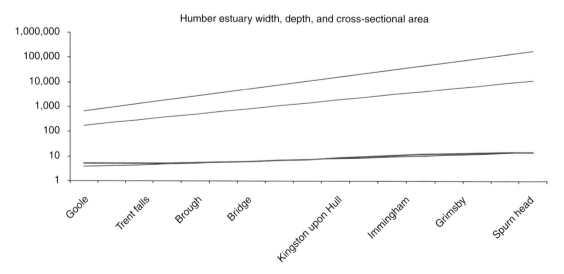

FIGURE 10.4 Longitudinal variation in temperature (pale line).

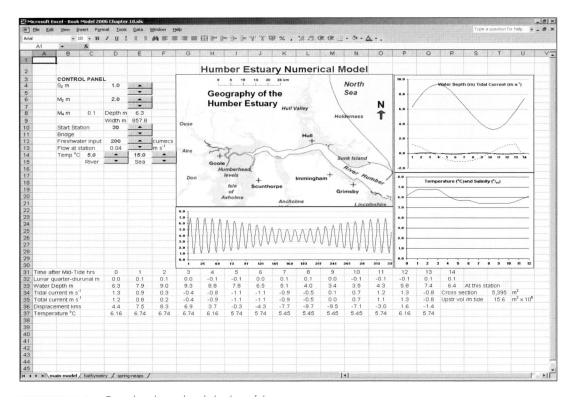

FIGURE 10.5 Completed graphical display of the estuarine temperatures.

water will be assumed to be fresh. There are only three stages in the work:

1 Select A38 on the bathymetry spreadsheet and enter Salinity ‰.
2 Select C38 and enter =35*C35 to use the Gaussian distribution to model the salinity distribution.
3 Select C38 and center with one decimal place then fill right to S38 to model the salinity distributions along the estuary as shown in Figure 10.6. For example the estuarine salinities reduce from 35 at Spurn Head, through 8.8 at Kingston upon Hull to 0.1 at Goole.

It is now necessary to advect the salt distribution landward and seaward along the estuary with the flooding and ebbing of the tide. It is assumed in this model

that the salinity distribution defined on the bathymetry spreadsheet represents conditions at mid tide on the flood. There are three stages in this work:

1 Select A38 on the main model spreadsheet and enter Salinity ‰.
2 Select D38 and enter

$$= \text{LOOKUP}(\$D\$10 + D36, \text{bathymetry!}$$
$$\$C\$27:\$S\$27, \text{bathymetry!}\$C\$38:\$S\$38).$$

3 Format D38 to center alignment and one decimal place and fill right to Q38.

The result, for the settings given earlier, is that the salinity at the Bridge rises from 4.1 to 6.1 as the more saline seawater floods upstream and then falls to 1.6 as the fresher river water is advected downstream by the ebbing tide.

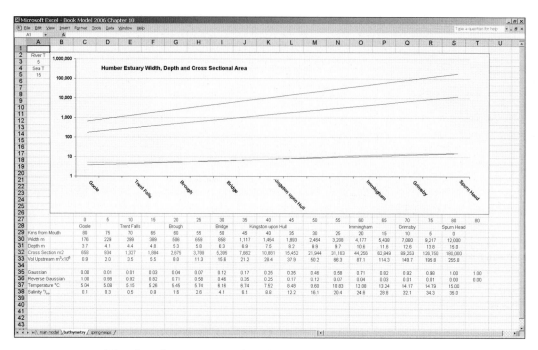

FIGURE 10.6   Modeling the salinity variations along the estuary.

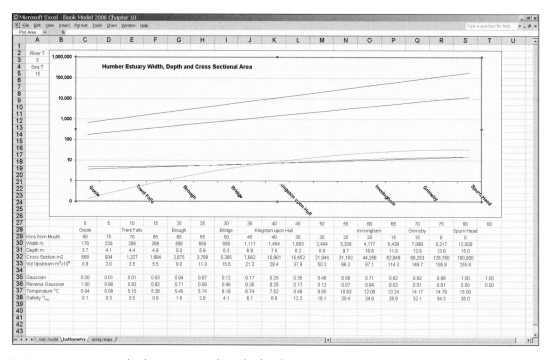

FIGURE 10.7   Longitudinal variation in salinity (broken line).

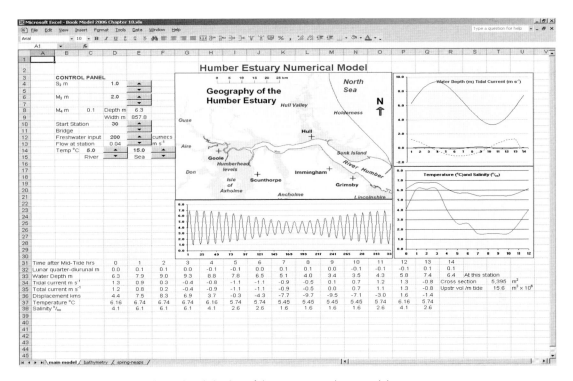

FIGURE 10.8 Completed graphical display of the estuarine salinity modeling.

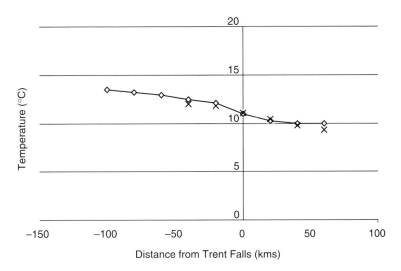

FIGURE 10.9 Comparison of Gameson (1982) temperature data (broken line and diamonds) with model results (crosses) for $M_2 = 2.0$, $S_2 = 1.0$, $Q = 200$ cumecs, and sea and river temperatures of 9 and 12°C respectively.

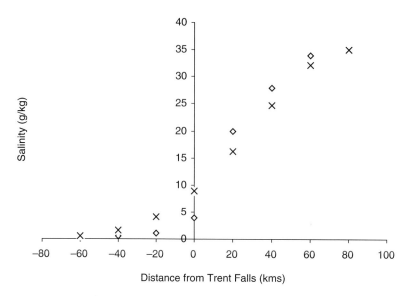

FIGURE 10.10   Comparison of Gameson (1982) salinity data (diamonds) with model results (crosses) for the same settings as in Figure 10.9 at high water.

10.7   DISPLAYING THE SALINITY DISTRIBUTION

The salinity distribution is displayed as both a longitudinal profile on the bathymetry sheet and as a new time series on the main model sheet:

1 Select the bathymetry sheet and click on the chart and then select the add data command in the chart dialog box.
2 Select the salinity data cells C38 to S38 and click ok.
3 Select the new line on the chart and format it appropriately (Figure 10.7).
4 Select the temperature chart on the main model spreadsheet and select add data from the chart menu.
5 Select D38:P38 to add the salinity data to the chart and click ok.

6 Finally, change the title of the chart and adjust the vertical scale as required to view the data as shown in Figure 10.8.

10.8   MODEL VALIDATION

The model's temperature and salinity variations are similar to those described in Chapter 1 for the Humber Estuary.

For example, there is a remarkably good comparison of observed and modeled temperatures between Goole and Kingston upon Hull with judicious choice of the model inputs as shown in Figure 10.9. The salinities also compare reasonably well, but could be calibrated by more careful choice of the dispersion coefficient as shown in Figure 10.10.

Part VI

SUSPENDED PARTICULATE MATTER IN ESTUARIES

11

ESTUARINE PARTICULATES

11.1 INTRODUCTION

The majority of the sediment which is interchanged between fluvial and marine systems is in the form of estuarine Suspended Particulate Matter (SPM). For example, Uncles et al. (2001) estimated that more than 90% of the 18–24×10^9 T of sediment delivered from rivers into the world's oceans each year is carried in suspension through estuaries. The concentration of the estuarine SPM depends not only upon the tidal range and mixing within the estuarine system but also varies through the tidal cycle and in response to seasonal freshwater inputs (Figure 11.1). Brown (1999) reports that SPM increases from approximately 10^2 mg dm^{-3} in estuaries with a small tidal range where turbulent mixing is generally weak, to approximately 10^4 mg dm^{-3} in estuaries with a large tidal range where turbulent mixing is stronger. This chapter considers the processes that suspend, transport, and deposit particulates in estuarine systems. There is a long history of research into the Humber's SPM (e.g. Jackson, 1964) and here, in particular, we derive functional relationships for the concentration and the location of particulates in order to develop the modeling work in the following chapter.

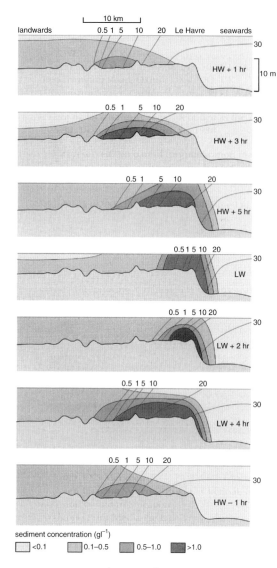

FIGURE 11.1   The particulate concentrations and movement of the estuarine turbidity maximum in the Seine Estuary, France, at hourly intervals during a spring tide cycle. Numbered lines are isohalines with salinity in parts per thousand. The diagram illustrates many of the processes which control estuarine SPM: the ebbing tide advects the particulates downstream and concentrations increase as erosion develops, before the turning tide brings material upstream once again (after Brown, 1999).

11.2   BACKGROUND INFORMATION

The suspension, transport, and deposition of particulate matter in tidal and estuarine systems is dealt with in texts such as Dyer (1986, 1997) and Masselink and Hughes (2003). This chapter is based, in particular, upon an interpretation of Markofsky et al. (1986), Brenon and Le Hir (1998), Clarke and Elliott (1998), and Tattersall et al. (2003). In general, the rate of change of the mass of SPM per unit area of the bed integrated through the water column (the mass of suspended load) depends upon advection, diffusion, erosion (source terms), and deposition (sink terms). These processes may be expressed by the depth-averaged advection–diffusion equation for suspended sediment (e.g. McManus and Prandle (1997) and Bass et al. (2002)). The suspended load is simply the product of the depth and the concentration, $h\overline{C}$:

$$\frac{\partial h\overline{C}}{\partial t} = u\frac{\partial h\overline{C}}{\partial x} + \frac{\partial}{\partial x}\left(K_x\frac{\partial h\overline{C}}{\partial x}\right) + E_r - D_p$$

Eq. 11.1

where $\overline{C}$ is the depth-averaged concentration, h is the depth, K_x is the horizontal diffusion coefficient, and u is the velocity component. The first term on the right-hand side represents material advected into the region by horizontal concentration gradients and is described below. The second term on the right-hand side represents horizontal diffusion and is generally neglected because the horizontal concentration gradients are small as is also described below. The last two terms on the right-hand side are the sources (the erosion) and sinks (the deposition) of sediment.

11.3   EROSION OF PARTICULATES

Solid particles are raised and transported downstream by fluid forces which balance the grain weight against gravity. The fluid forces are derived from the shear of the flow (Chapter 7). The problem resolves into the determination of the value of the fluid stress which initiates this movement and then the quantification of the particulate profiles in terms of the flow variables. The simplest determination of the shearing stress which the flow exerts on the sediment is known as the quadratic stress law (Lewis, 1997):

$$\tau = \rho C_{D100} u_{100}^2 \qquad \text{Eq. 11.2}$$

where τ is the shear stress (N m^{-2}), ρ is the seawater density (about 1000 kg m^{-3}), u_{100} is the tidal current speed at 100 cm above the bed, and C_{D100} is the coefficient of proportionality known as the drag coefficient. This expression is valid for a flow in which the Reynolds number (Chapter 7) is sufficiently high for friction to depend upon the roughness of the sedimentary surface so that the influence of viscosity is negligible. Dyer (1986) quotes the values for C_{D100} shown in Table 11.1.

Typical examples of the shear stress in an estuarine channel are shown in Figure 11.2 from Eq. 11.2 giving a maximum of about 10 N m^{-2} for a flow of 2 m s^{-1}.

The threshold of sediment transport is the critical value of the shear stress above which transport is initiated and, in general, the threshold increases with increasing grain

diameter. Dyer (1986) provides the data shown in Figure 11.3 for the critical condition as a function of the flow velocity. The data are represented here by:

$$U_{100cr} = 10.5 D^{0.37} \qquad \text{Eq. 11.3}$$

where D is the grain diameter. Eq. 11.3 is functionally similar to, for example, Miller et al. (1977) except that the multiplier there is 122.6 and the exponent is 0.29 for small particles and the grain diameter in millimeters.

The rate of erosion of particulates, E_p, is taken to be proportional to the excess bed shear stress (e.g. Dyer, 1986; Bass et al., 2002):

$$E_p = M \left(\frac{\tau}{\tau_{cr}} - 1 \right) \qquad \text{Eq. 11.4}$$

This formulation has, according to Dyer (1986), been used in mathematical modeling by Odd and Owen (1972) and Ariathurai and Krone (1976) with values for the erosion coefficient, M, in the range 0.005–0.015 kg m^{-2} s^{-1}, being greater with higher temperatures. More recently Brenon and Le Hir (1998) utilized $M = 0.001$ kg m^{-2} s^{-1}, Uncles et al. (1992) used $M = 0.00003$ kg m^{-2} s^{-1}, and Tattersall et al. (2003) used $M = 0.000035$ kg m^{-2} s^{-1}. Substituting Eq. 11.2 into Eq. 11.4 yields a simple expression for the erosion rate with units of kg m^{-2} s^{-1}:

$$E_p = M \left[\frac{u^2}{u_{cr}^2} - 1 \right] \qquad \text{Eq. 11.5}$$

TABLE 11.1 Drag coefficients for typical seabed types, based on a velocity measured 100 cm above the bed (after Soulsby, 1983; Dyer, 1986).

Bottom type	C_{D100}	Bottom type	C_{D100}
Silt/sand	0.0014	Mud	0.0022
Sand/shell	0.0024	Sand/gravel	0.0024
Mud/sand/ gravel	0.0024	Unrippled sand	0.0026
Mud/sand	0.0030	Gravel	0.0047
Rippled sand	0.0061		

mins0.16pt The result of Eq. 11.5 for $M = 0.0001$ with the tidal current shown in Figure 11.2 and threshold current, u_{cr}, of 0.2 (silt), 0.4 (fine sand), and 0.8 (medium sand) m s^{-1} is shown in Figure 11.4. The

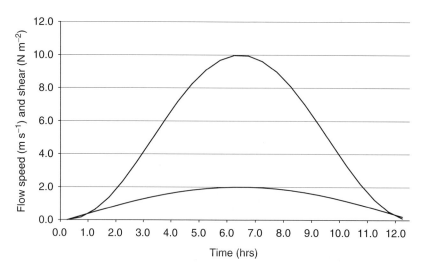

FIGURE 11.2 Typical results for the bed stress (upper line) from Eq. 11.2 for a tidal current (lower line) peaking at $2\,m\,s^{-1}$, a drag coefficient of 0.0025, and a water density of $1000\,kg\,m^{-3}$ (Toolbox 9).

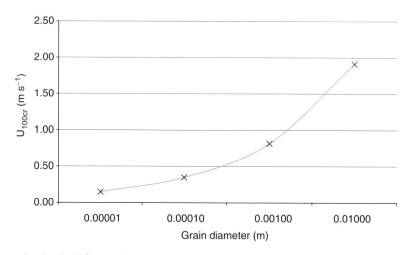

FIGURE 11.3 The threshold flow velocity on a flat bed for a current measured at a height of 100 cm. Data from Dyer (1986) after Miller et al. (1977) shown as crosses with the regression equation given here as Eq. 11.3 (Toolbox 10).

total mass eroded during a 12-hour tidal cycle in these cases was calculated as 219, 52, and $10\,kg\,m^{-2}$, respectively, which is equivalent to erosion of the surface by approximately 22, 5, and 1 cm respectively. These are reasonable values.

11.4 DEPOSITION OF PARTICULATES

The particles are continually settling under gravity. The fluid force F_D on a settling sphere can be written (Dyer, 1986) as

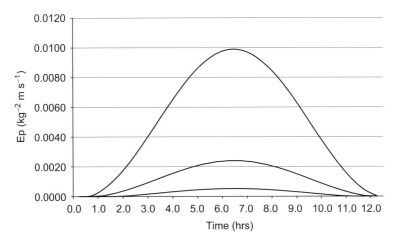

FIGURE 11.4 Erosion rates E_p from Eq. 11.5 for threshold currents of 0.2 (upper line), 0.4, and 0.8 m s^{-1} (lower line) corresponding to very fine to medium sand and an erosion coefficient, M, of 10^{-4} kg m^{-2} s^{-1} for the tidal current shown in Figure 11.2 (Toolbox 11).

follows:

$$F_D = C_D \pi \frac{D^2}{4} \rho \frac{w_s}{2} \qquad \text{Eq. 11.6}$$

where C_D is the sphere's drag coefficient, D is again the grain diameter, ρ is the density of the water, and w_s is the settling velocity. This force is balanced by the buoyancy force I:

$$I = \frac{4}{3}\pi \frac{D^3}{8}(\rho_s - \rho)g \qquad \text{Eq. 11.7}$$

At low Reynolds numbers, where the drag coefficient depends simply on the viscosity of the water and the particle size, and with quartz particles in sea water at a temperature of about 15°C and salinity of 35‰, these two expressions can be combined and simplified (Dyer, 1986) to the following:

$$\omega_s = 6000D^2 \qquad \text{Eq. 11.8}$$

Eq. 11.8 is known as Stokes Law and appears to apply reasonably well for particles up to medium sand. Typical values are in the range 0.03–3 mm s^{-1} (0.00003–0.003 m s^{-1} and Figure 11.5). The rate of deposition of

the particles D_p is then simply (e.g. Brenon and Le Hir, 1998; Bass et al., 2002)

$$D_p = \omega_s c_0 \qquad \text{Eq. 11.9}$$

Eq. 11.9 is based upon the so-called Krone formulation. In this expression c_0 is the concentration of suspended particulate matter close to the bed which some workers (e.g. Brenon and Le Hir, 1998) equate to the depth-averaged concentration. It is better, however, to recognize that c_0 depends upon the profile of particulates:

$$c_z = c_0 \left(\frac{h-z}{z} \cdot \frac{a}{h-a} \right)^{-B} \qquad \text{Eq. 11.10}$$

where, c_z is the particulate concentration at a height z above the bed, a is the reference height, h is the water depth, and B is the Rouse number given by, for example, Bass et al. (2002):

$$B = \frac{\omega_s}{\beta \kappa \sqrt{C_D} U} \qquad \text{Eq. 11.11}$$

β is the proportionality coefficient between eddy viscosity and diffusivity and is usually

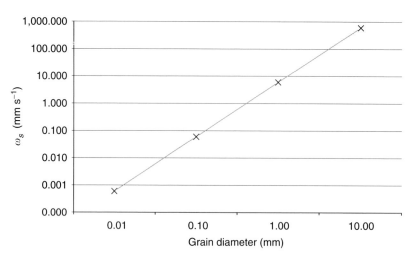

FIGURE 11.5 Stokes Law settling velocities (mm s^{-1}) as a function of grain diameter (mm) from Eq. 11.8 (Toolbox 12).

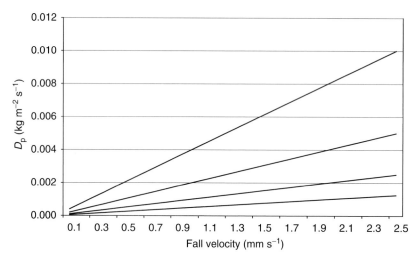

FIGURE 11.6 Deposition rates D_p from Eq. 11.12 for a concentration of 500 mg dm^{-3}, suspension parameters of 1 (lower line), 2, 4, and 8 (upper line) and a range of particle fall velocities (Toolbox 13).

taken as unity, and κ is von Karman's constant taken as 0.41. For the present purposes, the deposition rate is simply written as (after, e.g. Lumborg, 2002):

$$D_p = s_p \omega_s \overline{C} \qquad \text{Eq. 11.12}$$

where $\overline{C}$ is the depth average concentration and s_p, the suspension parameter, is the ratio $c_0/\overline{C}$ of the reference concentration to the depth-averaged concentration. The results for a typical concentration of 500 mg dm^{-3} and for a range of suspension parameters of 1, 2, 4, and 8 are shown in Figure 11.6 against a range of particle fall speeds. The total mass deposited each hour was calculated in these examples as 72, 360, and 720 kg m^{-2} for fall velocities of 0.1, 0.5, and 1 mm s^{-1}, respectively. These values are reasonable and compare with the erosion rates determined in Section 11.3.

The behavior of s_p is similar to the Rouse number with a low value indicating a well-mixed profile usually corresponding to higher velocity flows and lower fall speeds whereas the higher values indicate clearer water above a denser suspension which corresponds to lower velocity flows and higher fall speeds.

11.5 EQUILIBRIUM CONCENTRATIONS

The observation that, in tidally average terms, the rate of erosion must approximately balance the rate of deposition permits an independent estimate of some of the parameters which have been introduced above. For example, at equilibrium and in a symmetrical tidal current, so that the appropriate flood and ebb current speeds can be represented by a single mean u_m m s^{-1}, Eq. 11.5 and Eq. 11.12 must balance as follows:

$$C_{\max} = \frac{M}{s_p \omega_s} \left(\frac{u_m^2}{u_{cr}^2} - 1 \right)$$

Eq. 11.13

This expression is implemented in Toolbox 14 and the results are shown in Figure 11.7. It is apparent that some of the earlier estimates of the erosion coefficient M (Section 11.3), are too large. If, for example, $M = 0.01$ kg m^{-2} s^{-1} is used as indicated by Odd and Owen (1972) and Ariathurai and Krone (1976) along with a reasonable threshold velocity of 0.2 m s^{-1} for a well-mixed profile with $s_p = 2$ and a fall velocity of $\omega_s = 0.1$ mm s^{-1}, then a tidal current speed of only 1 m s^{-1} yields an SPM concentration in excess of 0.25×10^6 mg dm^{-3} (250 kg m^{-3}). More realistically, the value $M = 0.00003$ kg m^{-2} s^{-1} as indicated by Uncles et al. (1992) and Tattersall et al. (2003) yields SPM concentrations which are similar to those reported in the literature and in Section 11.1.

The results in Figure 11.7 show that these values of M give reasonable SPM concentrations. The concentration also decreases

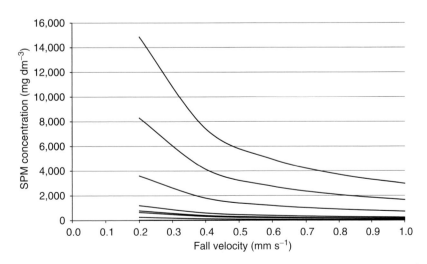

FIGURE 11.7 The SPM concentration at equilibrium as defined by Eq. 11.13 with $s_p = 2$, u_m from 0.5 to 2.0 m s^{-1} (lower to upper in each group) and for $M = 0.00002$ (lower dark lines) and 0.00006 (upper light lines) (Toolbox 14).

with the sediment fall velocity from, for example, about $16,000\,\mathrm{mg\,dm^{-3}}$ for $\omega_s = 0.2$–$3,500\,\mathrm{mg\,dm^{-3}}$ for $\omega_s = 1.0\,\mathrm{mm\,s^{-1}}$ and for $u_{cr} = 0.2\,\mathrm{m\,s^{-1}}$. Stronger tides increase the equilibrium concentration as discussed in greater detail below.

11.6   THE TURBIDITY MAXIMUM

Estuarine turbidity maximas (ETMs), in which the estuarine SPM peaks in the vicinity of the limit of salt intrusion (Figures 11.1 and 11.8), occur in many well-mixed and partially mixed estuaries (Allen et al., 1977; Officer and Nichols, 1980; Officer, 1981; Uncles and Stephens, 1993; Wolanski et al., 1995, 2003). For example, ETMs have recently been described in Chesapeake Bay by Sandford et al. (2001); in the Hudson River by Geyer (2001); in the Weser by Grabbemann and Krause (2001); in King

Sound, Western Australia by Wolanski and Spagnol (2003); and in the Loire by Ciffroy et al. (2003). ETM concentrations range from the order of $100\,\mathrm{mg\,dm^{-3}}$ in microtidal estuaries such as the Hawkesbury River, Australia to $20,000\,\mathrm{mg\,dm^{-3}}$ in macrotidal estuaries such as the Severn in the United Kingdom (Masselink and Hughes, 2003).

The ETM is maintained by a combination of one or more of three processes (Officer, 1981; Dyer, 1997):

1 **Vertical gravitational circulation.** SPM in the denser, near-bed landward flow converges with material in the lighter, surface seaward flow in stratified estuaries close to the limit of the salt intrusion (Figure 11.9).
2 **Tidal pumping.** Increasing tidal asymmetry due to the growth of the quarter-diurnal M_4 (Chapter 8) in well-mixed estuaries drives SPM shoreward on the flood until

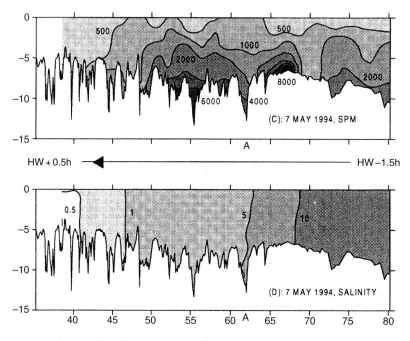

FIGURE 11.8   Longitudinal profiles of (top) SPM and (bottom) salinity in the Humber estuarine turbidity maximum (after Uncles et al., 1998).

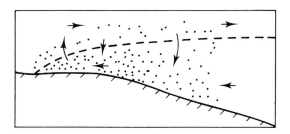

FIGURE 11.9 Estuarine turbidity maximum caused by residual circulation in a stratified (partially mixed or salt wedge) estuary with vertical exchanges across the level of no net motion (dashed line).

balanced by seaward directed, freshwater-enhanced transport on the ebb in the narrower, upper estuary and upstream of the tidal limit. The process is summarized in Figure 11.10 and considered in more detail in Section 11.9 below.

3 **Sediment dynamics.** Erosion and deposition lags occur because, although there is a threshold condition for the erosion of sediment, suspension and deposition can

continue even if the intensity of the flow drops below such threshold values. The result is a series of increasingly complex hystereses in SPM concentrations in estuarine environments (Figure 11.11).

Dyer (1997) finds that, in general, tidal pumping is the dominant process in well-mixed estuaries. Although tidal pumping is likely to generate the turbidity maxima in disequilibrium, it is shown below that, once an equilibrium is established there is no net sediment flux and the concentration profile is simply maintained by Eq. 11.13.

It is worth noting that, although each of these three mechanisms has the potential to create the observed ETM concentrations, each also results in net flux of sediment and is thus inappropriate for equilibrium conditions. Instead, at equilibrium the form of the solution detailed in 11.5 must be applied. In particular, there must exist an SPM distribution within which the dynamic

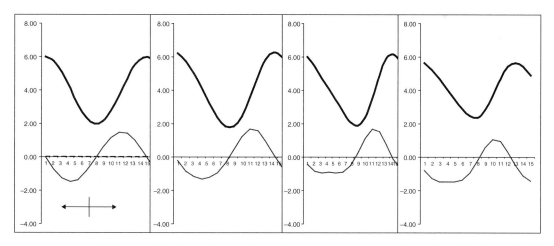

FIGURE 11.10 Schematic of surface elevation (upper lines) and currents (lower lines) from (left) the mouth of the estuary to (right) the head of the estuary. The surface elevation shows a range which decreases after initially increasing upstream and then becomes asymmetric with the growth of the quarter-diurnal harmonics. The flow is symmetrical at the mouth, becomes flood dominant with the increasing asymmetry, but then becomes ebb dominant as the freshwater discharge overcomes the tidal effects. The residual sediment transport is thus balanced at the mouth, is upstream in the lower reaches, and downstream above the null point (Toolbox 11.5.3).

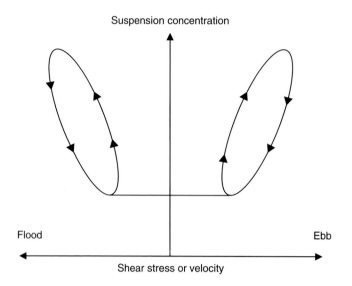

FIGURE 11.11 Schematic of erosion and sedimentation lags (after Masselink and Hughes, 2003).

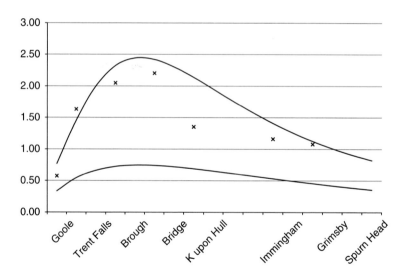

FIGURE 11.12 Comparison of the maximum current speeds from Eq. 11.14 (lower line) and the modeled equilibrium concentration from Eq. 11.15 (upper line) with the north bank SPM concentrations observed by R. Lewis on August 10, 2005 (crosses) (Toolbox 14).

equivalence between erosion and deposition is balanced throughout the estuary. To illustrate the point, consider the distribution of mid tide currents which arise from the flow modeling work in the previous chapters. The current speed is given by Eq. 7.18 as:

$$U(x, t) = \frac{\int_{x=X}^{x=L} W_0 \, e^{-ax/L} \Delta h_t \, dx - Q}{W_0 \, e^{-ax/L} D_0 \, e^{-bx/L}}$$

Eq. 11.14

and the corresponding particulate concentration is given by Eq. 11.13:

$$C_{max} = \frac{M}{s_p \omega_s} \left(\frac{u_m^2}{u_{cr}^2} - 1 \right) \qquad \text{Eq. 11.15}$$

Toolbox 14 includes an approximate evaluation of these distributions for the Humber with $M_2 = 2.0$, $S_2 = 1.0\,\text{m}$, $Q = 200\,\text{m}^3\,\text{s}^{-1}$, $M = 0.00003\,\text{kg}\,\text{m}^{-2}\,\text{s}^{-1}$, $s_p = 2.0$, $C_b = 100\,\text{mg}\,\text{dm}^{-3}$, $u_{cr} = 0.2\,\text{m}\,\text{s}^{-1}$, and $\omega_s = 0.1\,\text{mm}\,\text{s}^{-1}$. The results are shown in Figure 11.12.

There is clearly reasonable agreement between the equilibrium concentrations from Eq. 11.15 and the observed data. The equilibrium concentrations peak because the current speed increases upstream from the mouth as the cross-sectional area decreases more rapidly than the upstream volume until, at about Brough, the relative magnitudes are reversed and the current speed, and hence the equilibrium concentration, decreases.

11.7 INTRATIDAL FORCING OF SPM

Intratidal variations in the levels of SPM are due to erosion and deposition during the tidal cycle and to the advection of the ETM with respect to the observer.

Case 1: erosion and deposition

In general, concentrations rise when erosion dominates with the peak tidal currents at mid tide on both the flood and the ebb and then fall as deposition dominates at slack water for both high water and low water. The general relationship may be described, for example, by a cosinal function:

$$C(t) = \left(\frac{C_{max} - C_B}{2} \right) \left(1 + \cos \left(\frac{2\pi t}{6.21} \right) \right) + C_B$$

$$\text{Eq. 11.16}$$

where $C(t)$, C_{max}, and C_B are the concentration of SPM at time t, the maximum concentration, and the background concentration respectively. To a reasonable approximation, C_{max} may be taken to depend upon the ratio of the deposition and erosion rates from Eq. 11.4 and Eq. 11.12 as in Eq. 11.13. The result of combining Eq. 11.13 and Eq. 11.16 is implemented in Toolbox 14 and is shown in Figure 11.13. The amplitude of the intratidal variation increases if deposition dominates but decreases if erosion dominates, but overall the maximum value remains constant and the minimum value does not fall below background levels.

SPM variations due to advection of the ETM

Intratidal variations in the levels of SPM in response to movements of the ETM are summarized by Dyer (1986, 1997) as "at the seaward end of the turbidity maximum the peak of suspended sediment concentration will appear close to low slack water because of the advection from upstream.... At the upper end of the ETM the reverse happens, with the maximum concentrations occurring near high water slack" (Figure 11.14a).

Data utilized by Grabbemann and Krause (2001) confirms this description for the Weser Estuary showing that the phasing of these intratidal variations depends upon the location of the observer. The processes are also well illustrated by data for the Seine shown in Figure 11.1. The results are reproduced in Figure 11.14b with Toolbox 15 using an offset concentration of $20\,\text{mg}\,\text{dm}^{-3}$, a displacement amplitude of 20 km, an estuarine dispersion of 18, and an initial concentration of $200\,\text{mg}\,\text{dm}^{-3}$. The same toolbox can be used to investigate intratidal variations in the two main cases:

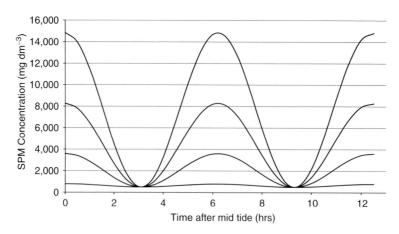

FIGURE 11.13   Intratidal variations in SPM due to erosion and deposition from Eq. 11.13 and Eq. 11.16 (Toolbox 14).

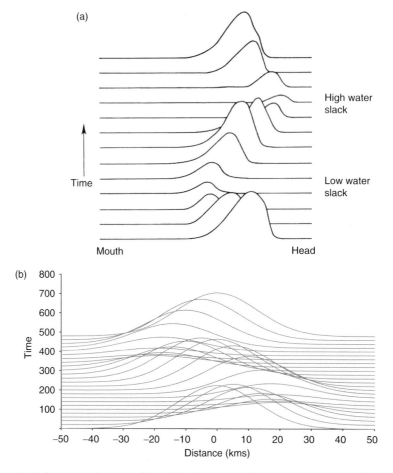

FIGURE 11.14   (a) Schematic advection of the ETM through a tidal cycle (after Dyer, 1986) and (b) numerical simulation based upon an advecting Gaussian distribution (Toolbox 14).

Case 2: observer upstream from ETM

In the upstream case (Figure 11.15) the ETM is advected away from the observer as the tide falls and SPM values fall, then toward the observer as the tide rises, although concentrations may reduce at high water slack. The water depth and SPM signatures are thus in phase.

Case 3: observer downstream from ETM

In the downstream case (Figure 11.15) the ETM is advected toward the observer as the tide falls and SPM values rise, then away from the observer as the tide rises, although concentrations may reduce at high water slack. The water depth and SPM signatures are thus out of phase.

11.8 INTERTIAL FORCING OF SPM

It is also well-known that the magnitude of the maximum concentration of SPM in the ETM increases with the tidal range (e.g. Tattersall et al., 2003). For example, Figure 11.16a shows Uncles et al.'s (1994) comparison of hourly-averaged SPM concentrations at 0.25 m above the bed at Halton Quay in the Tamar estuary, England with daily-averaged tidal range during November 1988. The near-bed levels fell to very low values during the neap tide on November 17, and subsequently increased on the rising spring tides.

There are two explanations which can account for the observed pattern and are here examined with Toolbox 16. The Toolbox simulates eight days of a neap-spring cycle through the addition of an M_2 and S_2 tide and displaces the water a distance given by

$$\Delta_{(t)} = -D_E(h_{(t)} - Z_0) \qquad \text{Eq. 11.17}$$

where $\Delta_{(t)}$ is the displacement due to a tide of depth $h_{(t)}$ about a mean sea level of Z_0 m and D_E is a coefficient.

Case 1: ETM advection

Figure 11.16(b) shows a simulation with Toolbox 16 using $Z_0 = 4$, $M_2 = 1.7$, $C_{\max} =$

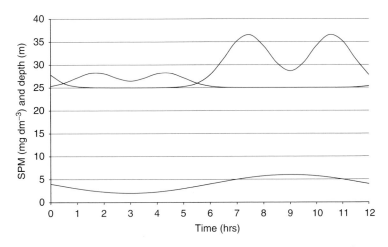

FIGURE 11.15 Numerical simulation based upon an advecting Gaussian distribution for an observer upstream of the ETM (top line) and downstream of the ETM (middle line) through the tidal depths (lower line) (Toolbox 15).

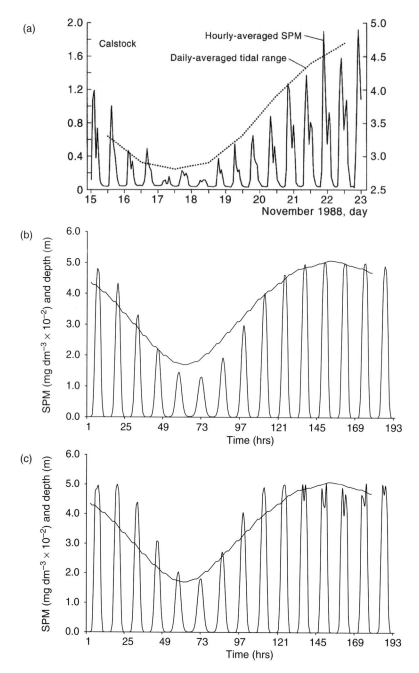

FIGURE 11.16 (a) Comparison of hourly-averaged SPM concentrations at Halton Quay in the Tamar Estuary, England with the daily-averaged tidal range for November 1988 (after Uncles, 1994). (b) and (c) Numerical simulations with parameters described in the text (Toolbox 16).

$500 \, \text{mg} \, \text{dm}^{-3}$, $D_E = 4 \, \text{km}$, $\sigma_p = 4$ for a station 10 km downstream of the ETM's mid tide location. The phenomena illustrated by Uncles' data in 11.16(b) is well reproduced and are attributable to the advection of the peak closer to the observer on the larger tides due to the stronger currents. The example can also be used to explain the double peaks in Uncles' data by simply increasing the tidal excursion from $D_E = 4$ to $D_E = 5 \, \text{km}$. The double peak is due to advection of the peak ETM past the observer generating their first peak and then back again for the second.

Case 2: erosion versus deposition

An alternative explanation for the relationship between tidal range and SPM levels depends upon the form of the erosion function. For example, Eq. 11.5 shows that the erosion rate depends, *inter alia*, upon the ratio u^2/u_{cr}^2 which increases with the tidal current speed and thus erosion and SPM concentration but deposition does not increase during the larger tidal ranges of spring tides.

11.9 SEASONAL FORCING OF SPM

In general, ETMs are forced downstream during periods of high freshwater discharge and upstream during periods of low freshwater discharge (Figure 11.17) and appear to coincide with a salinity value of about 2–5‰. For example, Uncles et al. (1998) report empirical relationships for the location and peak concentration of the Humber's ETM as a function of the tidal range and river discharge. The horizontal distance, x, has its origin, X_{ETM}, at the location of the turbidity maximum, which has been shown to depend in a simple fashion upon the freshwater input and the tidal range. Again for the

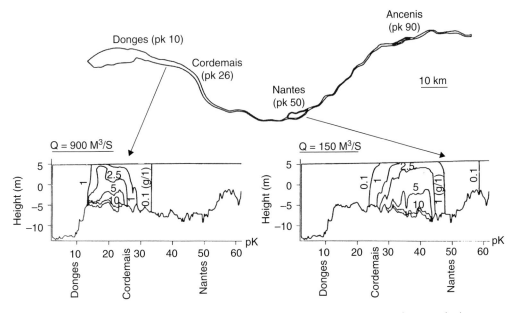

FIGURE 11.17 The movement upstream of the ETM in the Loire Estuary, France as the river discharge decreased from 900 cumecs to 150 cumecs (after Le Normant et al., 1998).

Humber:

$$X_{\text{ETM}} = 36 \log(Q_{31}) - 3.6\,R - 35$$
Eq. 11.18

where Q_{31} is the 31-day freshwater discharge (m^3s^{-1}) and R is the tidal range (m) (Uncles et al., 1998). The ETM is normally located about 50 km downstream from Naburn Lock, close to the confluence of the Ouse and the Trent, but varies from about 25–30 km downstream following the drought of the summer of 1995 to some 90 km downstream, in the vicinity of Kingston upon Hull, following the spates of February 1995. Tidal effects were shown to shift the turbidity maximum up-estuary by about 25 km between small neaps and large springs.

11.10 SUMMARY

This chapter has introduced the concepts of sediment erosion, deposition, advection, and diffusion in estuarine environments and derived theoretical and functional relationships for estuarine SPM concentrations and for the location and magnitude of the ETM.

The location of the ETM is given empirically for the Humber as follows:

$$X_{\text{ETM}} = 36 \log(Q_{31}) - 3.6R - 35 \quad \text{Eq. 11.19}$$

where Q_{31} is the 31-day freshwater discharge (m^3s^{-1}) and R is the tidal range (m) (Uncles et al., 1998). The ETM is thus driven upstream by larger tides and downstream by larger freshwater flows. The concentration of sediment at equilibrium may be calculated from

$$C_{\text{max}} = \frac{M}{s_{\text{p}}\omega_{\text{s}}}\left(\frac{u_{\text{m}}^2}{u_{\text{cr}}^2} - 1\right) \qquad \text{Eq. 11.20}$$

$$C(t) = \left(\frac{C_{\text{max}} - C_{\text{B}}}{2}\right)\left(1 + \cos\left(\frac{2\pi t}{6.21}\right)\right) + C_{\text{B}}$$
Eq. 11.21

where C_{max}, C_{B}, and $C(t)$ are the maximum, background, and time dependent SPM concentrations at time t ($\text{kg\,m}^{-3}\text{s}^{-1}$), M is the erosion coefficient ($M \approx 0.00003\,\text{kg\,m}^{-2}\,\text{s}^{-1}$), s_{p} is the suspension parameter ($s_{\text{p}} \approx 1$ for a well-mixed profile to $s_{\text{p}} \approx 8$ for a stratified profile), ω_{s} is the settling velocity ($\omega_{\text{s}} \approx 0.0001$ to $\omega_{\text{s}} \approx 0.003\,\text{m\,s}^{-1}$), and u_{cr} is the threshold flow velocity (in the range $u_{\text{cr}} \approx 0.1$–$u_{\text{cr}} \approx 1\,\text{m\,s}^{-1}$).

12

MODELING PARTICULATES

Contents

12.1 INTRODUCTION

Chapter 11 covered the erosion, deposition, advection, and diffusion of particulates in estuarine environments. The equilibrium SPM concentrations were discussed and the effects of tidal range and river discharge were examined. In essence the stronger the currents the greater the concentrations of SPM within the ETM. Large tidal ranges tend to move the ETM upstream, whilst large freshwater discharges tend to move the ETM downstream. The ETM itself also advects upstream during the flood and downstream during the ebb. In this chapter, the equilibrium profile is used to represent the diffusive distribution of particulates in the estuarine environment. Algorithms are incorporated that erode and deposit material and advect the distributions through the tidal cycle.

12.2 BACKGROUND INFORMATION

The preceding chapter demonstrated that, under equilibrium conditions, the erosion and deposition rates may approximately be equated to solve for the representative concentration:

$$C_{max} = \frac{M}{s_p \omega_s}\left(\frac{u_m^2}{u_{cr}^2} - 1\right) + C_B \qquad \text{Eq. 12.1}$$

where u_m and u_{cr} are the appropriate flow speed and threshold flow speed (m s^{-1}) and the units are mg dm^{-3} s^{-1}. The erosion coefficient M has a value of 0.0003 kg m^{-2} s^{-1}, suspension parameter s_p is 2 for a well-mixed profile, and the sediment fall velocity ω_s is in the range $0.1-1.0 \, \text{mm s}^{-1}$ ($0.0001-0.001 \, \text{m s}^{-1}$).

Separately, and to a first approximation, the peak tidal current depends upon the upstream volume, the freshwater discharge, and the cross-sectional area:

$$u_m = \frac{\text{upstream volume/hour}}{3600 * \text{width} \times \text{depth}} \quad \text{Eq. 12.2}$$

Finally, the intratidal changes in SPM may be represented by a cosinal function:

$$C(t) = \left(\frac{C_{max} - C_B}{2}\right)\left(1 + \cos\left(\frac{2\pi t}{6.21}\right)\right) + C_B$$
$$\text{Eq. 12.3}$$

These algorithms are incorporated into the model to simulate the behavior of the particulate matter.

12.3 SETTING UP PARTICULATES

In this section we add particulates to the control panel. There are only four stages in this work:

1 Open the model in Chapter 10 and "save as" with a new version number.
2 On the "main model" worksheet,

select	B17	and enter	PARTICULATES
	B18		M
	B19		s_p
	C18		0.00003
	C19		2
	D18		$\text{kg m}^{-2}\text{s}^{-1}$
	E18		Kms
	B20		C_b
	B22		u_{cr}
	B24		w_s
	C21		mg dm^{-3}
	C23		m s^{-1}
	C25		mm s^{-1}

3 Copy a spinner into each of the double cells as follows:

	Current	Max	Min	Increment	Cell Ref
D20:D21	10	10000	10	10	C20
D22:D23	1	100	1	1	D22
D24:D25	1	1000	1	1	D24

4 Select C22 and enter =D22/10
C24 =D24/10

Set $C_B = 100$, $u_{cr} = 0.2$, and $w_s = 0.2$ as shown in Figure 12.1.

12.4 DESCRIBING THE TURBIDITY MAXIMUM

In this section, we set up the distribution of suspended particulate matter in the estuary at mid tide on the flood and then erode and deposit this distribution, without advection:

1 On the bathymetry sheet

Select	A6	and enter	M
	A8		s_p
	A10		C_B
	A12		u_{cr}
	A14		w_s
	A7		= 'main model'!C18
	A9		= 'main model'!C19
	A11		= 'main model'!C20
A13		= 'main model'!C22
A15		= 'main model'!C24
A39		maximum flow m s^{-1}
A16		Tide
A17		= 2*('main model'!D4 + 'main model'!D6)

Center align A6:A15 and this should display the parameters as 0.00003, 2, 100, 0.2, and 0.2 respectively.

2 Select B41 to B53, center align the cells and enter 0 to 12 respectively.
3 Select C40 and enter =C27, then center align with zero decimal places and fill right to S27.
4 Select C39 and enter Eq. 12.2

$$= \left(\frac{0.15 * \$A\$17 * 1000000 * C33}{(3600 * C\$30 * C\$31)}\right)$$

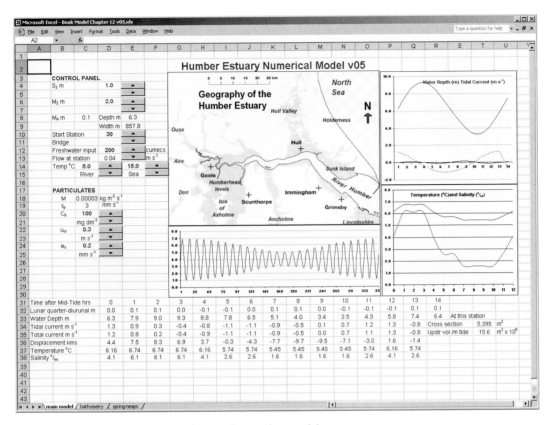

FIGURE 12.1 Preparing the particulates in the Humber model.

where 0.15 is one-sixth of the six-hour tidal semi-cycle, $A17 is the tidal range, 1000000 converts the volume into m³, C33 is the upstream tidal volume per meter of tidal rise, 3600 converts the hour into seconds, C$30 is the width at the station, and C$31 is the depth at the station.

5 Set C39 to center alignment with two decimal places and fill right to S39. The results with the inputs given earlier are shown in Figure 12.2 displaying a peak current at mid tide that increases from 0.33 m s⁻¹ at Goole to 0.75 m s⁻¹ at Brough and falling to 0.36 m s⁻¹ at Spurn Head.

12.5 CALIBRATING THE TURBIDITY MAXIMUM

The ETM suffers erosion and deposition throughout the tidal cycle. We first calculate these changes in concentration distribution with reference to the peak and then advect the evolving distribution in the following section.

1 Select C41 and enter Eq. 12.1

$$= \frac{1000*\$A\$7*(((C\$39\char94 2)/(\$A\$13\char94 2))-1)}{(\$A\$9*0.001*\$A\$15)+\$A\$11}$$

where 1000 converts kg m⁻³ into mg dm⁻³, $A$7 is the erosion coefficient M, C$39

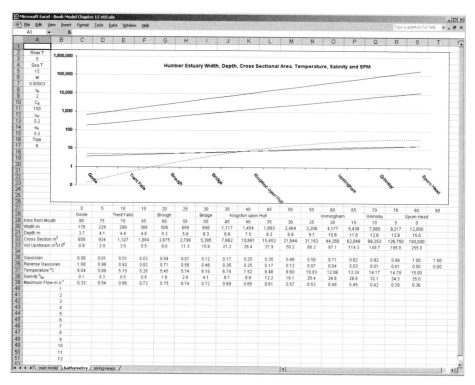

FIGURE 12.2   Entering the particulate data onto the bathymetry sheet.

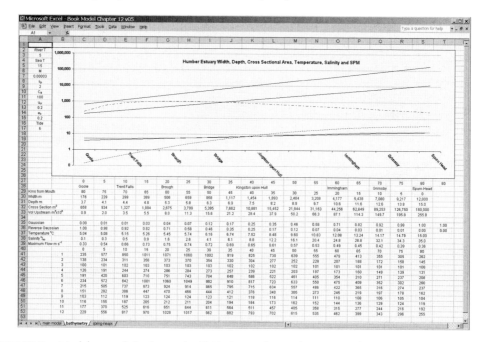

FIGURE 12.3   Modeling and displaying the SPM data without advection on the bathymetry sheet.

is the maximum station current speed $m\,s^{-1}$, A13$ is the threshold current speed $m\,s^{-1}$, A9$ is the suspension coefficient, 0.001 converts $mm\,s^{-1}$ into $m\,s^{-1}$, A15$ is the sediment fall velocity $mm\,s^{-1}$, and A11$ is the background concentration $mg\,dm^{-3}$.

2 Select C41 and center align with zero decimal places. Fill right to S41.

3 Select C42 and enter Eq. 12.2

$$=0.5*(C\$41-\$A\$11)$$
$$*(1+COS(26*PI()*\$B42/6.21))+\$A\$11$$

where C$41 is the maximum concentration at this station, $A$11 is background concentration, $B42 is the time in hours, and 6.21 is the period of the SPM cycles.

4 Set C42 to center alignment with zero decimal places and fill right and down to S53. The results with the inputs given earlier are shown in Figure 12.3 displaying an SPM

at, for example, Kingston upon Hull which, without advection, cycles from a maximum of 817 down to 102 and up again twice within the tidal cycle.

5 Add the data for the first hour of the tidal cycle in C41 to S41 to the display and amend the title accordingly as shown in Figure 12.3.

12.6 ADVECTING THE DISTRIBUTIONS

The mid tide distribution is advected upstream through the flood and then downstream through the ebb:

1 Select A39 on the main model spreadsheet and enter SPM Conc $mg\,dm^{-3}$.

2 Select D39 and enter

$$=HLOOKUP(\$D\$10+D36+5,$$
$$bathymetry!\$C40:\$S53,D31+2)$$

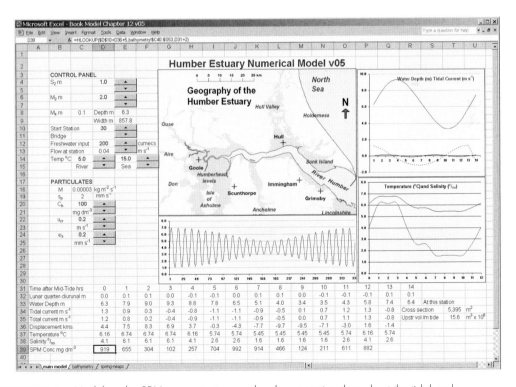

FIGURE 12.4 Modeling the SPM concentrations at the chosen station throughout the tidal cycle.

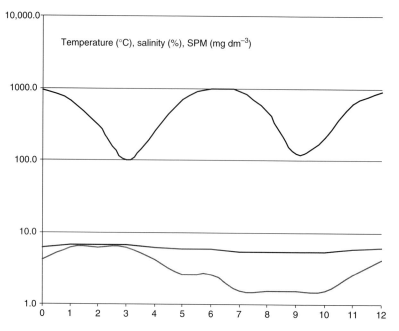

FIGURE 12.5 The particulate concentrations through the tidal cycle superimposed on the temperature and salinity graph.

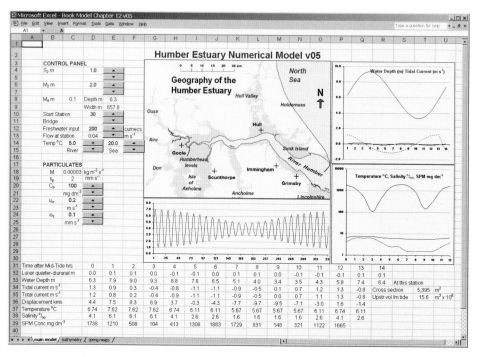

FIGURE 12.6 Completed numerical model of the Humber Estuary showing variation in water depth, current speed, temperature, salinity, and particulate concentrations.

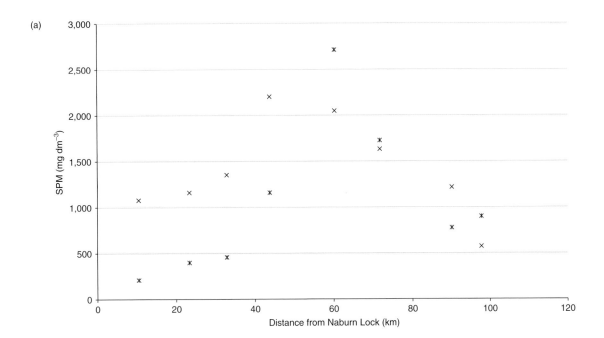

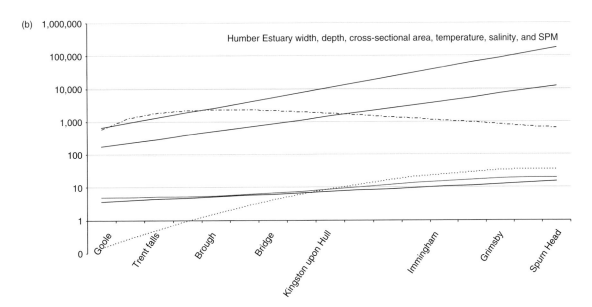

FIGURE 12.7 (a) Long profile variations in SPM from Naburn Lock to Spurn Head from high water, north bank surveys in August 2005 (data provided by R. Lewis) and (b) model simulation (broken line) showing peak SPM of about 2,500 mg dm^{-3} at Brough, approximately 60 km from Naburn Lock.

where the function looks up a site $D$10+D36+5 km downstream, $D$10 is the station, and D36+5 is the displacement (advection).

In the SPM array where D31+3 is the time step row reference.

3 Select D39, center to integer format and fill right to P39 as shown in Figure 12.4.

4 With the settings given, the SPM at Station 30 which is downstream of the ETM at mid tide, drops off from 919 to a minimum of 102 at high water then rises as the ETM advects towards the site on the ebb, reaching a maximum of 992 before dropping off again as the tide falls to low water.

12.7   GRAPHICAL DISPLAY OF PARTICULATES

1 Select the temperature and salinity chart on the main model.

2 Select add data from the Chart Menu and enter D39:P39.

3 Double click on the vertical axis and then select logarithmic display and set the maximum to 10,000 and the minimum to 1.

4 The resulting plot is shown in Figure 12.5.

5 The full model is now completed and is shown in Figure 12.6.

12.8   MODEL VALIDATION

This chapter has enabled the modeling of particulate erosion, deposition, advection, and diffusion and the results can be considered in terms of the longitudinal profile and in terms of the five temporal variations derived and discussed in Chapters 1–11. The model reasonably simulates the long profile (Figures 12.7a and 12.7b), it reasonably simulates the advection of the ETM and intratidal variations due to settling, and finally, it reasonably simulates longer-term spring-neap and seasonal variations in response to tidal range and freshwater flows.

BIBLIOGRAPHY

Acheson, D.J., 1990. *Elementary Fluid Dynamics*. Oxford, Oxford University Press, 397pp.

Allen, G.P., G. Sauzay, P. Castaing, and J.M. Jouanneau, 1977. Transport and deposition of suspended sediment in the Gironde Estuary, France. In: M. Wiley (ed.), *Estuarine Processes*. Academic Press, New York, pp. 63–81.

Ariathurai, R. and R.B. Krone, 1976. Finite element model for cohesive sediment transport. *J. Hydraul. Div., ASCE*, **102**, 323–38.

Bai, Y., Z. Wang, and H. Shen, 2003. Three-dimensional modelling of sediment transport and the effects of dredging in the Haihe Estuary. *Estuar. Coast. Shelf Sci.*, **56**, 175–86.

Bass, S.J., J.N. Aldridge, I.N. McCave, and C.E. Vincent, 2002. Phase relationships between fine sediment suspensions and tidal currents in coastal seas. *J. Geophys. Res.*, **107(C10)**, 1–14.

Bergman, T., 1784. *Physical and Chemical Essays* (translated by Edmund Cullen). Murray, London, 2 vol.

Birch, T. (ed.), 1965. *The Works of Robert Boyle*. Georg Olms, Hildeschiem, 6 vol.

Bloesch, J. and N.M. Burns, 1980. A critical review of sedimentation trap techniques. *Schweiz. Z. Hydrol.*, **42**, 15–55.

Boumans, R.M.J., D.M. Burdick, and M. Dionne, 2002. Modelling habitat change in salt marshes after tidal restoration. *Restor. Ecol.*, **10**, 543.

Boyle, R., 1693. An account of the honourable Robert Boyle's way of examining waters as to freshness and saltness. *Phil. Trans. R. Soc. Lond.*, **17**, 627–41.

Brenon, I. and P. Le Hir, 1998. Modelling fine sediment dynamics in the Seine estuary: interaction between turbidity patterns and sediment balance. In: J. Dronkers and M.B.A.M. Scheffers (eds.), *Physics of Estuaries and Coastal Seas*. Balkema, Rotterdam, pp. 103–14.

British Transport Board, 1977. *River Humber: Spurn Head to Trent Falls*. BTDB, Kingston upon Hull.

Brown, E., 1999. *Waves, Tides and Shallow-Water Processes*, 2nd edn. Butterworth-Heinemann, Oxford, 227pp.

Canada, 2004. Significant events in the history of hydrography. www.dfo-mpo.gc.ca/media/backgrou/2004/chs_e.htm

Chantler, A.G. 1974. The applicability of regime theory to tidal watercourses. *J. Hydraul. Res.*, **12(2)**, 181–92.

Chappell, J., 1993. Contrasting Holocene sedimentatry geologies of lower Daly River, northern Australia, and lower Sepik-Ramu, Papua New Guinea. *Sediment. Geol.*, **83**, 339–58.

Chappell, J. and C.D. Woodroffe, 1994. Macrotidal estuaries. In: R.G. Carter and C.D. Woodroffe (eds.), *Coastal Evolution: Late Quaternary Shoreline Morphodynamics*. Cambridge University Press, Cambridge, pp. 187–218.

Ciffroy, P., J.-L. Reyss, and F. Siclet, 2003. Determination of residence time of suspended particles in the turbidity maximum of the Loire estuary by [7]Be analysis. *Estuar. Coast. Shelf Sci.*, **57**, 553–68.

Clarke, S. and A.Q.J. Elliott, 1998. Modelling suspended sediment concentrations in the Firth of Forth. *Estuar. Coast. Shelf Sci.*, **47**, 235–50.

Coats, R., M.S. Kelly Kuff, and P. Williams, 1995. Using geomorphological relationships in designing tidal slough channels. *Proc. National Interagency Workshop on Wetlands: April 1995*. New Orleans, Louisiana.

Cox, C.A., 1963. The salinity problem. *Prog. Oceanogr.*, **1,** 243–61.

Dalrymple, R.W., R.J. Knight, B.A. Zaitlin, and G.V. Middleton, 1990. Dynamic and facies model of a macrotidal sand-bar complex, Cobequid Bay-Salmon River estuary (Bay of Fundy). *Sedimentology*, **37**, 577–612.

Dalrymple, R.W., B.A. Zaitlin, and R. Boyd, 1992. Estuarine facies models: conceptual basis and stratigraphic implications. *J. Sediment. Petrol.*, **62**, 1130–46.

Deacon, M.B., 1971. *Scientists and the Sea, 1650–1900*. Academic Press, London.

Delft Hydraulics, 1980. *Investigation Vlissingen Bottle*. M1710, Delft, The Netherlands.

Dittmar, W., 1884. *Physics and Chemistry*, Reptort on the scientific results of the voyage of H.M.S. Challenger during the years 1873–76, 1, 251pp.

Doodson, A.T. and H.D. Warburg, 1941. *Admiralty Manual of Tides*. HMSO, London, 270pp.

Dronkers, J., 2005. *Dynamics of Coastal Systems*. Advanced Series in Ocean Engineering – Vol. 25. World Scientific, Singapore, 519pp.

Dyer, K.R., 1986. *Coastal and Estuarine Sediment Dynamics*. Wiley, Chichester, 342pp.

Dyer, K.R., 1997. *Estuaries: A Physical Introduction*. Wiley, Chichester, 195pp.

Dyer, K.R. and A.L. New, 1986. Intermittency in estuarine mixing. In: D.A. Wolfe (ed.), *Estuarine Variability*. Academic Press, Orlando, pp. 321–39.

Dyer, K.R., M.-C. Robinson, and D.A. Huntley, 2001. Suspended sediment transport in the Humber Estuary. In: D.A. Huntley, G. Leeks, and D. Walling (eds.), *Land–Ocean Interaction:*

Measuring and Modelling Fluxes from River Basins to Coastal Areas. IWA Publishing, London, pp. 169–84.

Emery, W.J. and R.E. Thomson, 1998. *Data Analysis Methods in Physical Oceanography*. Pergamon Press, New York, 634pp.

Forchhammer G., 1865. On the composition of seawater in different parts of the ocean. *Phil. Trans. R. Soc. Lond.*, **155**, 203–62.

Gameson, A.L.H., 1982. Physical characteristics. In: A.L.H. Gameson (ed.), *The Quality of the Humber Estuary, A Review of the Results of Monitoring 1961–1981*. Humber Estuary Committee, Yorkshine Water Authority, Leeds, pp. 5–14.

Gao, S. and M.B. Collins, 1994. Tidal inlet equilibrium in relation to cross-sectional area and sediment transport patterns. *Estuar. Coast. Shelf Sci.*. **38**, 157–72.

Geyer, R.W., J.D. Woodruff, and P. Traykovski, 2001. Sediment transport and trapping in the Hudson Estuary. *Estuaries*, **24**, 670–9.

Grabemann, I. and G. Krause, 2001. On different time scales of suspended matter maximum in the Weser Estuary. *Estuaries*, **24**, 688–98.

Hamon, B.V., 1955. A temperature–salinity–depth recorder. Conseil permanent international pour le exploration de la Mer. *J. Conseil*, **21**, 22–73.

Hansen, D.V. and M. Rattray, 1966. New dimensions in estuary classification. *Limnol. Oceanogr.*, **11**, 319–26.

Hardisty, J., 1990a. *The British Seas: An Introduction to the Oceanography and Resources of the North-West European Continental Shelf*. Routledge, London, 272pp.

Hardisty, J., 1990b. *Beaches: Form and Process*. Unwin-Hyman, London, 324pp.

Hardisty, J., 2004. *Sediment Flux in the Humber Estuary*. Unpublished Report for the Environment Agency, 160pp.

Hardisty, J. and H.L. Rouse, 1996. The Humber Observatory: monitoring, modelling and management for the coastal environment. *J. Coast. Res.*, **12**, 683–90.

Hardisty, J., D. Taylor, and S.E. Metcalfe, 1993. *Computerised Environmental Modelling*. Wiley, Chichester, 276pp.

Hardisty, J., R. Middleton, D. Whyatt, and H.L. Rouse, 1998. Geomorphological and hydrodynamic results from digital terrain models of the Humber Estuary. In: S.N. Lane, K.S. Richards, and J.H. Chandler (eds.), *Landform Monitoring, Modelling and Analysis*. Wiley, Chichester, pp. 422–33.

Harvey, H.W., 1955. *The Chemistry and Fertility of Sea Water*. Cambridge University Press, Cambridge, 224pp.

Hay, A.E., 1983. On the remote acoustic detection of suspended sediment at long wavelengths. *J. Geophys. Res.*, **88(C12)**, 7525–42.

Hess, F.R. and K.W. Bedford, 1985. Acoustic Backscatter System (ABSS): the instrument and some preliminary results. *Mar. Geol.*, **66**, 357–80.

Hill, P.S. and I.N. McCave, 2001. Suspended particle transport in benthic boundary layers. In: B.P. Boudreau and B.B. Jorgensen (eds.), *The Benthic Boundary Layer: Transport Processes and Biogeochemistry*. Oxford University Press, Oxford, pp. 18–103.

Hood, W.G., 2004. Indirect environmental effects of dikes on estuarine tidal channels: thinking outside of the dike for habitat restoration and monitoring. *Estuaries*, **27**, 273–82.

Hume, T.M. and C.E. Herdendorf, 1992. Factors controlling tidal inlet characteristics on low drift coasts. *J. Coast. Res.*, **8**, 355–75.

Huntley, D.A., 1979. Tides on the North West European continental shelf. In: F.T. Banner, M.B. Collins, and K.S. Massie (eds.), *The North West European Shelf Seas: The Sea Bed and Sea in Motion, Vol. II. Physical and Chemical Oceanography, and Physical Resources*. Elsevier, Amsterdam, pp. 301–52.

Jackson, W.H., 1964. Effect of tidal range, temperature and fresh water on the amount of silt in suspension in an estuary. *Nature*, **201**, 1017.

Jarret, J.T., 1976. *Tidal Prism–Inlet Area Relationships*. GITI Rep. 3, US Army Eng. Waterways Sta., Vicksburg.

Kitheka, J.U., M. Obiero, and P. Nthenge, 2005. River discharge, sediment transport and exchange in the Tana estuary, Kenya. *Estuar. Coast. Shelf Sci.*, **63**, 455–68.

Knudsen, M., C. Forch, and S.P.L. Sorensen, 1902. Bericht uber die chemische und physikalische Untersuchung des Seewassers und die Aufstellung der neuen hydrographischen Tabellen Wiss. *Meeresunters*, n.f., **6**, 13–184.

Krause, N.C., 1987. Application of portable traps for obtaining point measurements of sediment transport in the surf zone. *J. Coast. Res.*, **3**, 2.

Krumbein, W.C., 1934. Size frequency distributions of sediments. *J. Sedim. Petrol.*, **4**, 65–77.

Lane, S.N., J.H. Chandler, and K.S. Richards, 1994. Developments in monitoring and modelling small-scale river bed topography. *Earth Surf. Proc. Land.*, **19**, 349–68.

Lavoisier, A., 1772. Memoire sur l'usage de esprit-de-vin dans l'analyse des eaux minerales, *Mem. Acad. R. Sci. (Paris)*, 555–63.

Le Normant, C., 2000. Three-dimensional Modelling of Cohesive Sediment transport in the Loire estuary. *Hydrological Processes*, **14**, 2231–43.

Lewis, R., 1997. *Dispersion in Estuaries and Coastal Waters*. Wiley, Chichester, 312pp.

Linklater, E., 1972. The Voyage of the *Challenger*. Cardinal, London, 288pp.

Long, A.J., J.B. Innes, J.R. Kirby et al., 1998. Holocene sea-level change and coastal evolution in the Humber Estuary, eastern England: an assessment of rapid coastal change. *The Holocene*, **8**, 229–47.

Lumborg, U., 2002. Cohesive sediment transport modelling – application to the Lister Dyb tidal area in the Danish Wadden Sea. *Jnl. Coast. Res.*, **41**, 114–23.

Lynch, J.F., 1985. Theoretical analysis of ABSS data for HEBBLE. *Mar. Geol.*, **66**, 277–89.

Mackay, H.M. and E.H. Schumann, 1990. Mixing in Sundays River estuary, South Africa. *Estuar. Coast. Shelf Sci.*, **31**, 203–16.

McManus, J.P. and D. Prandle, 1997. Development of a model to reproduce observed suspended sediment distributions in the southern North Sea using principal component analysis and multiple linear regression. *Cont. Shelf. Res.*, **17**, 761–78.

Mallowney, B.M., 1982. Mathematical modelling. In: A.L.H. Gameson (ed.), *The Quality of the Humber Estuary, A Review of the Results of Monitoring 1961–1981.* Humber Estuary Committee, Yorkshire Water Authority, Leeds, pp. 82–90.

Markofsky, M., G. Lang, and R. Schubert, 1986. Suspended Sediment Transport in Rivers and Estuaries. In: van de Kreeke (ed.), *Lecture Notes on Coastal and Estuarine Studies Nr. 16: Physics of Shallow Estuaries and Bays.* Springer, New York, 210–227.

Masselink, G. and M.G. Hughes, 2003. *Introduction to Coastal Geomorphology and Processes.* Arnold, London, 354pp.

Massey, B.S., 1970. *Mechanics of Fluids.* Van Nostrand Reinhold, London, 508pp.

Metcalfe, S.E., S. Ellis, B.P. Horton, et al. 2000. The Holocene evolution of the Humber Estuary: reconstructing change in a dynamic environment. In: I. Shennan and J. Andrews (eds.), *Holocene Land–Ocean Interaction and Environmental Change around the North Sea.* Geological Society, London, Special Publications, **166**, pp. 97–118.

Miller, M.C., I.N. McCave, and P.D. Komar, 1977. Threshold of sediment motion under uni-directional currents. *Sedimentology,* **24**, 507–27.

Millero, F.J., P. Chetirkin, and F. Culkin, 1977. The relative conductivity and density of standard seawaters. *Deep-Sea Res.,* **24**, 315–21.

Moore, I.D., R.B. Grayson, and A.R. Ladson, 1991. Digital terrain modelling: a review of hydrological, geomorphological and biological applications. *Hydrol. Proc.,* **5**, 3–30.

Neumann, G. and W.J. Pierson, 1966. *Principles of Physical Oceanography.* Prentice-Hall, Englewood Cliffs, NJ, 545pp.

O'Brien, M.P., 1969. Equilibrium flow areas of inlets and sandy coasts. *J. Waterways, Harbour and Coast. Eng. Div.,* **95**, 43–52.

Odd, N.V.M. and M.W. Owen, 1972. A two layer model of mud transport in the Thames Estuary. *Proc. Inst. Civil Eng.,* Supplement 9. Paper 75175.

Officer, C.B., 1976. *Physical Oceanography of Estuaries and Associated Coastal Waters.* Wiley, Chichester, 465pp.

Officer, C.B., 1981. Physical dynamics of estuarine suspended sediments. *Mar. Geol.,* **40**, 1–14.

Officer, C.B. and M.N. Nichols, 1980. Box model application to a study of suspended sediment distribution and fluxes in partially mixed estuaries. In: V. Kennedy (ed.), *Estuarine Perspectives.* Academic Press. London, pp. 329–40.

Owen, M.W., 1975. *Erosion of Avonmouth muds.* Hydraulics Research Station Report INT 150.

Petrie, G. and T.J.M. Kennie, 1987. Terrain modelling in surveying and civil engineering. *Comput. Aided Des.,* **19**, 171–87.

Poisson, A.T., T. Dauphinee, C.K. Ross, and F. Culkin, 1978. The reliability of standard seawater as an electrical conductivity standard. *Oceanol. Acta,* **1**, 425–33.

Prandle, D., 1985. On salinity regimes and the vertical structure of residual flow in narrow tidal estuaries. *Estuar. Coast. Shelf Sci.,* **20**, 615–35.

Prandle, D., 1986. Generalised theory of estuarine dynamics. In : van de Kreeke (ed.), *Physics of Shallow Estuaries and Bays.* Springer-Verlag, Berlin, pp. 42–57.

Psuty, N.P. and M.E.S.A. Moreira, 2000. Holocene sedimentation and sea level rise in the Sado estuary, Portugal. *J. Coast. Res.,* **16**, 125–38.

Pugh, D.T., 1987. *Tides, Surges and Mean Sea Level.* Wiley, Chichester, England.

Pugh, D.T., 2004. *Changing Sea Levels. Effects of Tides, Weather and Climate.* Cambridge University Press, Cambridge. 265pp.

Rees, J.G., J. Ridgway, S. Ellis, R.W. O'B. Know, R. Newsham, and A. Parkes, 2000. Holocene sediment storage in the Humber Estuary. In: I. Shennan and J. Andrews (eds.), *Holocene Land–Ocean Interaction and Environmental Change around the North Sea.* Geological Society, London, Special Publications, **166**, 119–44.

Ridout, P., 2004. *The Laboratory Measurement of Salinity.* www.seawatersolutions.com

Riley, J.P. and G. Skirrow, 1965. *Chemical Oceanography*, Vols. I and II. Academic Press, New York. 712pp.

Roy, P.S., 1984. New South Wales estuaries: their origin and evolution. In: B.G. Thom (ed.), *Coastal Geomorphology in Australia*. Academic Press, Sydney, pp. 99–121.

Roy, P.S., P.J. Cowell, M.A. Ferland, and B.G. Thom, 1994. Wave-dominated coasts. In: R.G. Carter and C.D. Woodroffe (eds.), *Coastal Evolution: Late Quaternary Shoreline Morphodynamics*. Cambridge University Press, Cambridge, pp. 121–86.

Ruddiman, W.F., M.E. Raymo, D.G. Martinson, B. Clement, and J. Backman, 1989. Pleistocene evolution: Northern hemisphere ice sheets and North Atlantic ocean. *Palaeoceanography*, **4**, 353–412.

Sandford, L.P., S.E. Suttles, and J.P. Halka, 2001. Reconsidering the physics of the Chesapeake Bay estuarine turbidity maximum. *Estuaries*, **24**, 655–69.

Semeniuk, V., 1985. Mangrove environments of Port Darwin, Northern Territory: the physical framework and habitats. *J. R. Soc. Western Aust.*, **68**, 53–79.

Shennan, I. and J.E. Andrews, 2000. An introduction to Holocene land-ocean interaction and environmental change around the western North Sea. In: I. Shennan and J. Andrews (eds.), *Holocene Land–Ocean Interaction and Environmental Change around the North Sea*. Geological Society, London, Special Publications, **166**, pp. 1–8.

Shennan, I., W.R. Peltier, R. Drummond, and B. Horton, 2002. Global to local scale parameters determining relative sea-level changes and the post-glacial isostatic adjustment of Great Britain. *Quat. Sci. Rev.*, **21**, 397–408.

Simmons, H.B., 1955. Some effect of upland discharge on estuarine hydraulics. *Proc. Am. Soc. Civ. Eng.*, **81**, 792.

Somoza, L. and J. Rey, 1991. Holocene fan deltas in a "ria" morphology: prograding clinoform types and sea-level control. *Cuadernos Geol. Iberica*, **15**, 37–48.

Soulsby, R.L., 1983. The bottom boundary layer of shelf seas. In: B. Johns (ed.), *Physical Oceanography of Coastal and Shelf Seas*. Elsevier, Amsterdam. Chapter 5.

Sternberg, R.W., 1986. Transport and accumulation of river-derived sediment on the Washington Continental Shelf, USA. *Journal of the Geological Society London*, **143**, 945–56.

Summerfield, M.A., 1991. *Global Geomorphology*. Longmans, Harlow, England, 537pp.

Sverdrup, H.V., M.W. Johnson, and R.H. Fleming, 1942. *The Oceans*. Prentice Hall, New York.

Syvitski, J.P.M., D.C. Burrell, and J.M. Skei, 1987. *Fjords: Processes and Products*. Springer-Verlag. New York, 379pp.

Tattersall, G.R., A.J. Elliott, and N.M. Lynn, 2003. Suspended sediment concentrations in the Tamar estuary. *Estuar. Coast. Shelf Sci.*, **57**, 679–88.

Thain, R.H., A.D. Priestley, and M.A. Davidson, 2004. The formation of a tidal intrusion front at the mouth of a macrotidal, partially mixed estuary: a field study of the Dart estuary, U.K. *Estuar. Coast. Shelf Sci.*, **61**, 161–72.

Thorne, P.D., C.E. Vincent, P.J. Hardcastle, S. Rehman, and N. Pearson, 1991. Measuring suspended sediment concentrations using acoustic backscatter devices. *Mar. Geol.*, **98**, 7–16.

UKHO, 2004. *Admiralty TotalTide* (CD-ROM). United Kingdom Hydrographic Office, London.

UKHO, 2004. A brief history of the United Kingdom Hydrographic Office. www.ukho.gov.uk/corp/History.asp

Uncles, R.J. and J.A. Stephens, 1993. The fresh–saltwater interface and its relationship to the turbidity maximum in the Tamar Estuary, U.K. *Estuaries*, **16**, 126–41.

Uncles, R.J., A.J. Bale, R.J.M. Howland, A.W. Morris, and R.C.A. Elliott, 1983. Salinity of surface water in a partially-mixed estuary, and its dispersion at low run-off. *Oceanol. Acta*, **6**, 289–96.

Uncles, R.J., J.A. Stephens, and M.L. Barton, 1992. Observation of fine-sediment concentrations and transport in the turbidity maximum region of an estuary. In: D. Prandle (ed.), *Dynamics and Exchanges in Estuaries and the Coastal Zone. Coastal and Estuarine Studies*, Vol. 40. Berlin, Springe, pp. 255–76.

Uncles, R.J., J.A. Stephens, and M.L. Barton, 1994. Seasonal variability of fine-sediment flux in the Tamar Estuary. In: K.R. Dyer and R.J. Orth (eds.), *Changes in Fluxes in Estuaries: Implications from Science to Management.* Olsen & Olsen, Frednsborg, Denmark, pp. 17–22.

Uncles, R.J., A.E. Easton, M.L. Griffiths, et al., 1998. Seasonality of the turbidity maximum in the Humber-Ouse estuary, U.K. *Mar. Pollut. Bull.,* **3**, 206–15.

Uncles, R.J., S.J. Lavender, and J.A. Stephens, 2001. Remotely sensed observations of the turbidity maximum in the highly turbid Humber Estuary, U.K. *Estuaries,* **24**, 745–55.

Van de Noort, R., and S. Ellis, 1997. *Wetland Heritage of the Humberhead Levels; an Archaeological Survey.* Humber Wetlands Project.

Van Rijn, L.C., 1990. *Principles of Fluid Flow and Surface Waves in Rivers, Estuaries and Coastal Seas.* Aqua Publications, Amsterdam.

Van Rijn, L.C., 1993. *Principles of Sediment Transport in Rivers, Estuaries and Coastal Seas.* Aqua Publications. Amsterdam.

Vincent, C.E., D.M. Hanes, and A.J. Bowen, 1991. Acoustic measurement of suspended sand on the shoreface and the control of concentration by bed roughness. *Mar. Geol.,* **96**, 1–18.

Wells, J.T., 1995. Tide-dominated estuaries and tidal rivers. In: G.M.E. Perillo (ed.), *Geomorphology and Sedimentology of Estuaries.* Elsevier, Amsterdam, pp. 179–205.

White, M., 1999. Measuring water clarity from space. *The Irish Scientist Yearbook 1999.* Oldbury Publishing, Dublin.

Williams, D.J.A., and J.R. West, 1975. Salinity distribution in the Tay Estuary. *Proceedings of the Royal Society of Edinburgh,* **75B**, 29–39. Edinburgh.

Willgoose, G.R. and G.R. Hancock, 1998. Revisiting the hypsometric curve as an indicator of form and process in transport-limited catchment. *Earth Surf. Proc. Land.,* **23(7)**, 611–23.

Wolanski E. and R.J. Gibbs, 1995. Flocculation of suspended sediment in the Fly river estuary. Papua New Guinea. *J. Coast. Res.,* **11(3)**, 754–62.

Wolanski, E., and S. Spagnol, 2003. Dynamics of the turbidity maximum in King Sound, tropical Western Australia. *Estuar. Coast. Shelf Sci.,* **56**, 877–90.

Woodroffe, C.D., 2003. *Coasts: Form, Process and Evolution.* Cambridge University Press, Cambridge, 623pp.

Wright, L.D., J.M. Coleman, and B.G. Thom, 1973. Processes of channel development in a high-tide range environment: Cambridge Gulf-Ord River Delta, Western Australia. *J. Geol.,* **81**, 1.

Young, R.A., J.T. Merrill, T.L. Clarke, and J.R. Proni, 1982. Acoustic profiling of suspended sediments in the marine bottom boundary layer. *Geophys. Res. Lett.,* **9**, 17.

Index